Amazon Unfiltered

Power, Performance, and the Human Cost of Scale

Rad Stephens

For Becky, who lived every version of this book long before it was written. Each day, she made time to listen patiently as I shared my triumphs, struggles, frustrations, and small victories. She always knew when to ask the hard question, when to offer perspective, and when to simply let me finish the story.

And for Amanda, Cortney, Alec, and Emily. You are the reason any of it matters.

Amazon Unfiltered: Power, Performance, and the Human Cost of Scale

First edition

Published by Rad Stephens Media LLC

This book is a work of nonfiction based on the author's professional experience, observations, research, and interpretation. Some names, identifying details, timelines, roles, and circumstances have been changed, combined, or compressed to protect privacy, preserve confidentiality, and focus attention on broader organizational patterns.

Composite characters and scenes are used where appropriate. They are intended to represent recurring conditions and observed patterns, not to identify specific private individuals.

The views expressed in this book are the author's own.

ISBN hardcover: 979-8-9965245-0-1

Table of Contents

BEFORE YOU ENTER THE MACHINE 1

PROLOGUE 4

ENTERING NIGHT SHIFT: MY FIRST LOOK INSIDE 7

BUILD FOR SCALE 11

THE LANGUAGE OF THE MACHINE 18

THE NUMBERS THAT RUN THE OPERATION 22

THE SYSTEMS BEHIND THE PROMISE 30

AMAZON LEADERSHIP PRINCIPLES 38

FROM CLICK TO DOORSTEP 42

INSIDE AMAZON'S FULFILLMENT CENTER 48

INSIDE AMAZON'S SORT CENTERS 54

INSIDE AMAZON'S DISTRIBUTION CENTERS 56

INTRODUCTION TO THE DSP PROGRAM 60

DSP OPERATIONAL CHALLENGES AND SOLUTIONS 64

DSP RESILIENCE 68

ON-ROAD WITH DAS 71

THE REALITY BEHIND OWNING YOUR OWN DSP BUSINESS 74

HELPING DSPS NAVIGATE A SYSTEM BUILT AGAINST THEM 79

FIRST DAYS INSIDE THE DSP AND DA WORLD 81

THE ROAD REVEALS EVERYTHING 83

DISPATCH CYCLES 88

SEEING DAS AS MORE THAN JUST WORKERS 90

THE DISPATCH MACHINE 92

HOW AMAZON FORECASTS CAPACITY 100

INSIDE THE RETURN TO STATION PROCESS 104

RTS PHYSICAL LAYOUT AND VEHICLE MANAGEMENT . 108

TURNING FRICTION INTO DATA 110
RTS QUALITY CONTROL AND DSP ACCOUNTABILITY 112
RTS STANDARD OF WORK 114
REDISCOVERING PURPOSE IN AN UNEXPECTED PLACE 118
MEETING THE MACHINE 121
NAVIGATING CULTURAL DYNAMICS AND BUILDING ALLIANCES 123
THE CHAOS OF HIGH VOLUME AND THE THREAT OF VIOLENCE 127
OBSERVING LEADERSHIP CHALLENGES 131
BREAKING POINTS 134
OPERATIONAL BLIND SPOTS NO ONE TALKS ABOUT 137
LEADERSHIP AND RECOGNITION 140
PORTLAND: A NEW JOURNEY 144
PORTLAND'S VIBRANT CULTURE AND UNFORGETTABLE FOOD 146
FROM ASSOCIATE TO LEADER 149
PERKS ON PAPER 152
A UNIQUE STATION AND SEASONAL CHALLENGES 154
LESSONS FROM REMARKABLE LEADERS 157
UNLIKELY MENTORS IN A SEA OF CHAOS 160
TALENT RECOGNIZED, POTENTIAL REALIZED 164
GRILLS, BONDS, AND JÄGER BOMBS 166
DSP BEGINNING LESSONS 168
PORTLAND: LESSONS FROM THE FIELD 172
THE BOTTLENECK BREAKTHROUGH 176
TRUST THROUGH ACTION 179
BUILDING CONNECTIONS BEYOND THE WORKPLACE . 182
EVALUATORS LACKED RELEVANT OPERATIONAL EXPERIENCE 184

LEAVING A LASTING LEGACY....188
TIME TO TRANSITION....191
SETTLING INTO MARWOOD HILLS....193
LEADERSHIP AND AMAZON'S CULTURAL SHIFT....195
RISING TO LEADERSHIP AND INVENTING....201
UNLIKELY LESSONS FROM DELIVERY SERVICE PROVIDERS....205
MANAGER'S BALANCING ACT....209
THE STRAIN ON ASSOCIATES AND THE PUSH FOR CHANGE....214
THE COMPLEXITY OF OPERATIONAL PROCESSES....220
EMPATHY IN THE WORKPLACE....223
A BITTERSWEET DEPARTURE....228
BULLYING BECAME NORMAL....231
WHEN THE WORK FINALLY FELT RIGHT....239
WORK HARD, PLAY HARD, DELIVER HARDER....243
THE PEOPLE WHO HELD THE MIDDLE....246
YOUR PROFIT MATTERS....250
AMAZON'S STRATEGIC MISSTEP....253
SAVING ROUTES IN REAL TIME....258
FINALLY IN THE DRIVER'S SEAT....262
SYSTEM FAILS PEOPLE....265
I TRAINED THEM, THE SYSTEM WEAPONIZED THEM....268
STRETCH ROLES AND BROKEN MARGINS....272
THE PANDEMIC....276
SENT WHERE EVERYTHING WAS BREAKING....281
PANDEMIC PROMOTION....286
THE PANDEMIC FALLOUT....289
USING REAL-TIME DATA TO FIX WHAT WAS BROKEN....296
NEW STATION LAUNCHES....303

MEGA STATION LAUNCH 308
HOW AMAZON DESTROYS ITS BEST PEOPLE 315
FIXING WHAT THE SO-CALLED EXPERT BROKE 326
PERSISTENT OPERATIONAL IMPROVEMENTS 335
PILLARS OF CONTINUOUS IMPROVEMENT 339
THE ONLY ONE WHO BELIEVED 348
LEADERSHIP WITHOUT LIVED EXPERIENCE 358
BUILT TO BE SAFE 365
RESPECT IS EARNED ON THE FLOOR 373
WHEN INTEGRITY TRIGGERS HOSTILITY AND BULLYING 381
HR WAS NEVER A SAFE PLACE 389
A NEW DEMON 398
TOXIC LEADERSHIP, PROTECTED POWER 408
FINALLY, A BREATH OF FRESH AIR 416
THE WORKFORCE AMAZON NEVER SAW 422
EFFICIENCY HAS A COST 434
THE ROLE THAT EXPLAINED EVERYTHING 439
INHERITED FAILURE 442
THIS IS WHAT LEADERSHIP LOOKED LIKE 447
STANDARD WORK WITHOUT ACCOUNTABILITY 452
WHEN THE WORK SPOKE FOR ITSELF 456
THE POINT OF NO RETURN 459
CONTROLLING THE NARRATIVE 462
WHAT IT COST, AND WHAT REMAINED 465
THE COST OF AMAZON'S FUTURE 468
EPILOGUE: TO THOSE STILL INSIDE THE MACHINE 471
AUTHOR'S NOTE 473
ABOUT THE AUTHOR 475

Before You Enter the Machine

The first time I realized something was wrong, it was not during a crisis, a missed metric, or an obvious failure. It was during a meeting that went exactly as planned. A room full of capable managers sat quietly while a senior leader walked through a slide deck everyone had already reviewed. The metrics were clean, the forecast was optimistic, and the risks had been acknowledged, then softened with language that sounded responsible without committing to anything real. When the leader asked if there were questions, no one spoke. I looked around the room and saw recognition on several faces, followed by the same calculation I was making myself. This was not the moment to push. This was the moment to align.

The meeting ended on time. Action items were assigned, the plan moved forward, and two weeks later, the operation absorbed the cost. The failure was reframed as an execution issue, attributed downward, and corrected just enough to keep the narrative intact. No one lied, no one broke the rules, and nothing about the system changed.

That moment stayed with me because it revealed something essential. This was not a company that failed because people were incompetent or malicious. It was a company that succeeded precisely because people learned when not to speak. The silence was not confusion, and it was not agreement. It was calculation, shaped by experience inside a system where challenging the approved direction often carried more risk than letting the operation absorb the consequences later.

This book is not written as a traditional memoir, nor is it a simple exposé. It is a guided walk through a system few people ever see in full, told by someone who lived inside it long enough to understand how it actually functions. The early chapters focus on orientation, explaining how Amazon's logistics machine is built, how its metrics operate, and how decisions made far from the warehouse floor shape daily reality. These chapters are intentionally detailed because they ground the reader in the mechanics of the operation before judgment is passed. Without that foundation, the human stories that follow would be easy to dismiss as isolated, exaggerated, or anecdotal, and they are not.

As the book progresses, the focus shifts from systems to people. The reader moves from process explanations into lived experience, from dispatch lanes and capacity models into leadership behavior, cultural incentives, and the quiet compromises required to survive inside a performance-obsessed environment. This transition is deliberate because the machine does not fail people all at once. It does so gradually, through incentives that reward compliance, silence, and optics over integrity and results.

The middle sections document growth, opportunity, and genuine pride alongside mounting tension. These chapters reflect the paradox at the heart of Amazon, where extraordinary innovation exists alongside deeply corrosive leadership practices. You will see moments of mentorship, teamwork, and operational brilliance, but you will also see how those moments are undermined by politics, fear-driven management, and systems that punish those who challenge broken processes.

The later chapters confront consequences. The pandemic, leadership failures, and internal accountability mechanisms expose how power is protected and how dissent is neutralized. These chapters are not written to shock. They are meant to show how institutional behavior becomes normalized, even when it causes real harm. What happens to one employee is rarely unique. It is usually a preview.

The chapters ahead do not spare the reader from what that degradation looks like at its worst. They describe a colleague who, from my perspective, weaponized an investigation to block careers and was later rewarded with promotion. They describe a manager who mocked an injured subordinate's medical shoes in front of the floor, only for the remark to disappear from the written record. They describe a workplace harassment investigation conducted by someone only months into their first corporate role, despite a record of documented incidents that deserved far more care and experience.

They describe a performance review built on meetings that allegedly occurred on the reviewer's days off and processes for which no training had ever existed, then used as pressure toward a resignation that never came. They also describe a meeting with a leader so fully captured by another person's narrative that he attacked work he did not appear to understand, delivered with the confidence of someone insulated from the consequences of being wrong.

These are not the book's lowest moments because they are dramatic. They are its lowest moments because none of them are surprising by the time they arrive. That is the point. The toxicity does not erupt all at once; it accumulates. Each chapter in this sequence becomes the predictable next step of a system that had been signaling its direction for years. By the time the decision to leave finally comes, it does not feel like defeat. It feels like the only rational response to information that had been building for a very long time.

This book does not ask the reader to hate Amazon. It asks the reader to see it clearly and to understand how a company capable of extraordinary execution can simultaneously erode trust, health, and humanity within its own walls. The story you are about to read is personal, but it is not singular. It reflects patterns that repeat wherever scale, data, and unchecked authority collide. Read it as a journey, not a verdict. The conclusions will make more sense once you see how they were built.

Prologue

"Through the Gears: My Journey Inside Amazon's Machine"

By the time I first stepped into Amazon's labyrinthine network, I already understood that systems do not reveal themselves all at once. They reveal themselves through repetition, through meetings that end cleanly, through metrics that look reasonable on a screen, and through decisions that feel small enough to accept in the moment. Still, walking into the beating heart of Amazon's logistics machine felt like entering the physical manifestation of modern ambition itself. This was no ordinary company. It was a colossal engine of innovation, where technology and logistics intertwined to move billions of packages to doorsteps across the globe, and where precision was not just admired, but demanded.

I was captivated by Amazon's ability to bend time and distance in the name of convenience. Almost immediately, another question followed. At what cost? That question became the quiet undercurrent of my career, threading through moments of pride, frustration, and eventual disillusionment. It never fully left me, no matter how far I advanced or how many problems I solved.

From the beginning, Amazon demanded constant adaptation. The pace was unforgiving. Every role required not only technical competence, but leadership, creativity, and resilience. I thrived on solving complex logistical problems and building tools that turned raw data into clarity. I helped design systems that exposed root causes, improved delivery performance, and made sense of chaos while the operation never stopped moving. Those wins were not just about hitting targets. They were about understanding how a massive machine actually worked and making it work better.

It would take time to understand that the same precision that made the system so effective at scale also determined who absorbed the strain when efficiency demanded a sacrifice.

But innovation came paired with relentless pressure. Amazon's obsession with metrics often eclipsed its concern for people. Beneath the technological excellence was a culture driven almost entirely by numbers. Completion rates. Defect metrics. Productivity benchmarks. There was little room for error and even less room for humanity. Over time, it became clear that people were often treated less like contributors and more like variables to be optimized.

There were leaders who inspired growth and modeled what good leadership could look like. They existed, and they mattered. But they were the exception. Far more common were leaders who governed through fear. Public takedowns during Weekly Business Reviews. Decisions wrapped in jargon that shut down dissent rather than encouraging solutions. What should have been collaborative problem-solving often became a performance of dominance. I watched capable people burn out, strong leaders sidelined by politics, and teams slowly eroded by a culture that made many question their own worth.

And yet, even inside that pressure cooker, there were moments of humanity that kept me going. I built deep bonds with colleagues who cared about doing the work right, even when the system worked against them. I found meaning in mentoring associates and managers, watching them gain confidence and grow into leaders themselves. Those moments grounded me. They reminded me that empathy and resilience still had a place, even inside a machine often designed to test both.

Amazon is a paradox. It represents the height of human ingenuity, a place where innovation reshapes what is possible every single day. It is also a machine that demands everything from the people inside it, often without acknowledging the toll. My journey reflects both truths. It is a story of growth and frustration, of awe at what technology can accomplish and anger at how easily people can be overlooked in the process.

This book is my attempt to tell the truth about what life inside Amazon really feels like, especially from the perspective of someone embedded in the grind of package distribution. I want you to feel the rhythm of the operation, the brilliance of the systems, the determination of the people who kept everything moving, and the darker edges of a culture that too often valued output over human cost. You will encounter leaders who inspired me and others who embodied exactly what happens when power goes unchecked.

This is not just my story. It is a look inside the engine room of a company that shapes the modern world and leaves a lasting mark, for better or worse, on everyone who steps inside it.

Welcome to Amazon, where every day is Day One. Where innovation collides with intensity, and where ambition and pressure move side by side. This is what it's like to live and work inside the machine.

To understand what ultimately breaks, the next chapters step back to show how the machine is constructed, how it governs behavior, and why its failures are rarely accidental.

Entering Night Shift: My First Look Inside

When I joined Amazon as a Seasonal Associate, I was thrust into the heart of a dynamic, high-energy team that played a pivotal role in the company's logistics chain. The Ashford Distribution Center wasn't just another warehouse; it was the crucial final stop before packages reached customers' doorsteps. This hub acted as a bridge between the sprawling network of Fulfillment and Sort Centers and the last-mile delivery process. Our job, though seemingly straightforward, was vital: taking the influx of packages from various upstream centers, sorting them meticulously by destination, and ensuring they were routed correctly for on-time delivery. Every package represented a promise, and the pressure to meet Amazon's high-speed delivery expectations was palpable.

One of the standout aspects of the role was the scheduling flexibility Amazon offered, which made the demanding nature of the job more manageable. Seasonal workers, in particular, benefitted from being able to choose shifts that aligned with their other commitments. Whether someone preferred overnight, early morning, daytime, evening, or weekend hours, there was a shift for everyone. For those looking for a different kind of balance, Amazon even offered options like four-day workweeks with three-day weekends. Amazon made every attempt to accommodate schedules, but most of the time you were assigned a shift. However, if the assigned shift did not work, you could request another shift, and usually, it was granted. This adaptability was a significant draw, providing a semblance of work-life balance in an environment often characterized by its relentless pace. It was a small yet impactful way for the company to acknowledge the diverse needs of its workforce.

I was assigned to the night shift, where my primary responsibilities revolved around receiving and preparing inventory for delivery. While much of the work involved physically moving packages, technology served as the backbone of the process, ensuring efficiency and precision. One of the core tasks I handled was the "Pick and Stow" process, which heavily relied on handheld scanning devices. Each shift, I would retrieve packages from the conveyor belt, scan their labels, and place them into the designated tote or bin for their delivery stop. Although it seemed straightforward, this process required immense speed and accuracy. A single error, whether a mislabeled item or a misplaced package, could disrupt the entire logistics chain, causing delays or complications down the line.

A significant part of the role also involved unloading enormous 53-foot linehaul trucks that arrived throughout the night. These trucks, packed floor-to-ceiling with parcels from fulfillment centers across the region, needed to be efficiently unloaded, sorted, and prepared for the next leg of their journey. One of the most physically demanding tasks, known as "Waterspider," required using pallet jacks to move gaylords, large containers holding over 300 pounds of packages, to specific locations within the warehouse. This task demanded not just physical strength but also precision and endurance, as navigating these heavy loads through the warehouse was both challenging and essential to maintaining the flow of operations. The physicality of the job added an intense layer of complexity, requiring methodical approaches to keep up with the relentless pace.

Over time, I gained experience performing all the roles on the night shift, including tasks like "Pick and Stage," waterspider, Divertor, and Problem-Solve. Each role brought its own challenges, but together they offered a comprehensive view of the intricate logistics operation. Mastering these diverse responsibilities deepened my appreciation for the complexity and coordination required to ensure that every package moved seamlessly through the system and ultimately reached its destination.

The night shift for me, running from 11 p.m. to 6 a.m. four days a week, came with its own set of challenges. Adjusting to a nocturnal schedule meant battling fatigue and finding ways to maintain focus during the long, physically demanding hours. Yet, despite these difficulties, the night shift developed its own rhythm. There was a distinct energy to working in the quiet hours when the rest of the world seemed to be sleeping. Over time, I grew to appreciate the stillness and routine of those nights, which gave the work a unique sense of purpose and camaraderie among the team.

After three months of dedication and hard work, I was promoted to a full-time associate, which came with extended hours. My new schedule was four 10-hour shifts, from 10 p.m. to 8:30 a.m., giving me three full days off each week. At first, I welcomed the longer shifts. Having three days off in a row seemed like the perfect opportunity to recover from the week's work, recharge, and enjoy some personal time. However, as I delved deeper into the full-time role, I realized that the physical exertion of unloading and sorting thousands of packages each night took a toll on my body.

The job was intense, and by the end of each week, I was often left exhausted. The first day of my break was usually spent recovering from the physical strain. My body needed time to heal from the constant lifting, moving, and sorting. By the second day, I would start to feel more like myself, but there was still a lingering sense of mental fatigue. It often took me until the third day to feel fully rested and clear-headed. As much as I enjoyed the longer weekends, I also realized that it came with the need for extensive recovery time.

My time as an associate was physically demanding and left little room for romanticizing the work. It required stamina, routine, and a constant negotiation between effort and recovery. The experience did not just build endurance. It revealed how quickly physical limits became part of the operational equation, shaping what people could realistically sustain over time.

Looking back, that first stretch as a Seasonal Associate stripped the work down to its most honest form. There were no abstractions, no dashboards, no leadership language to hide behind. It was just bodies, time, gravity, and volume, night after night. The job taught me early that Amazon's speed was not powered by technology alone, but by people pushing themselves to physical limits that were rarely acknowledged once the shift ended. That experience became the baseline for everything that followed. It grounded my understanding of the operation in reality, shaped how I evaluated decisions made far above the floor, and left me with a lasting respect for the people who carried the weight of the system long before anyone called it a career.

Build for Scale

Ranking fifth globally by market capitalization at approximately $2.49 trillion as of January 2026 (Google), Amazon has solidified its position as a transformative force in both commerce and technology. Starting as an online bookstore in Jeff Bezos' garage in 1994, Amazon quickly evolved into a multifaceted powerhouse, reshaping the way we shop, live, and even entertain ourselves.

Its reach now stretches far beyond e-commerce. What started as an online retailer has grown into a sprawling tech ecosystem that includes AWS, a leading force in cloud computing; a fast-rising digital advertising business; Prime Video and Amazon Studios; physical stores like Whole Foods and Amazon Fresh; and devices ranging from Echo and Fire TV to Kindle. Amazon also runs one of the largest logistics networks in the world, operates Twitch in the gaming space, owns Ring in home security, and has moved into healthcare through One Medical and Amazon Pharmacy, all supported by subscription services like Prime, Audible, and Kindle Unlimited. The company's ubiquitous influence has made it one of the most recognizable brands globally, to the extent that "Amazon" is increasingly used as a verb, much like "Google."

Despite its scale today, Amazon's origin story follows a familiar arc of ambition, timing, and relentless execution. Like other tech giants such as Google and Facebook, Amazon's meteoric rise began with humble beginnings, driven by innovation, strategic diversification, and relentless customer focus. Amazon's history is marked by pivotal milestones and strategies that propelled its evolution from a fledgling startup into a global conglomerate, continually redefining industries at scale.

Jeff Bezos attended Princeton University, where he majored in electrical engineering and computer science, graduating with honors in 1986. Following his education, he embarked on a career on Wall Street, holding positions at notable firms including Fitel, Bankers Trust, and D. E. Shaw & Co. At D. E. Shaw, a hedge fund renowned for its focus on advanced mathematical modeling, Bezos encountered a pivotal statistic: the internet was growing at a then-unprecedented rate of 2,300% annually. This discovery ignited his entrepreneurial vision and set the stage for one of the most transformative ventures in modern business history.

Recognizing the immense potential of the burgeoning internet, Bezos envisioned an online bookstore that could eventually expand to sell virtually anything. On July 5, 1994, he officially founded his company under the name "Cadabra," operating out of his garage in Bellevue, Washington. However, the name was quickly changed to "Amazon" after concerns that "Cadabra" sounded too similar to "cadaver," which could create negative associations. The new name was inspired by the vastness and grandeur of the Amazon River, symbolizing his ambition to build the world's largest online retailer.

By July 1995, Amazon officially launched its website as an online bookstore, offering customers across all 50 U.S. states and 45 countries access to its expanding catalog. Remarkably, much of this operation was still being run out of Jeff Bezos's modest garage workspace. That early launch was less about books and more about proving a concept. Amazon was built from the start around scale, speed, and customer obsession. Those early decisions laid the groundwork for what the company would eventually become, a global e-commerce force driven by relentless innovation and an uncompromising focus on growth and customer experience.

Like many startups, Amazon grappled with financial challenges in its early years. Despite these hurdles, the company demonstrated remarkable growth in 1996. Between the third and fourth quarters, Amazon's revenue surged from $4.2 million to $8.5 million, a significant milestone that coincided with a $100,000 reduction in losses. This progress marked a pivotal moment, suggesting that its business model was gaining traction and moving toward sustainability

(Wilhelm, 2019). While Amazon did not achieve profitability until the late 1990s, its unwavering focus on innovation and customer-centric strategies began to yield tangible results, setting the stage for its future dominance.

In 1998, Amazon began diversifying beyond books, marking the beginning of its expansion into a vast range of product categories. This strategic move laid the groundwork for Amazon's evolution into a comprehensive online marketplace. The company achieved its first profitable quarter in the fourth quarter of 2001, driven by a strong holiday shopping season (ABC News, n.d.).

By 2003, Amazon posted its first full year of profitability, reporting net earnings of $35 million after a staggering $149 million loss the year before (Hansell, 2004). The shift was more than a financial milestone. It validated a strategy that prioritized long-term growth over short-term comfort and proved the company could survive its own ambition. That first profitable year cemented Amazon's credibility, not just as an experiment in online retail, but as a durable business model capable of scaling at an unprecedented pace. It also marked the moment when expansion was no longer a gamble, but an expectation, setting the stage for the dominance that would follow.

From its modest start as a niche online bookseller, Amazon has grown into one of the most influential companies in the world. Its early financial struggles became a proving ground for its resilience, adaptability, and innovative strategies, ultimately establishing Amazon as a global leader across e-commerce, technology, and numerous other industries.

A major milestone in Amazon's evolution was the launch of Amazon Prime in February 2005. Initially introduced as a $79 annual subscription offering unlimited two-day delivery, Prime experienced slow adoption in its early years. At the time, online shopping was still an emerging trend, and many consumers were hesitant to commit to the seemingly high subscription cost. However, Amazon's long-term strategy of enhancing Prime's value gradually turned the tide.

Key improvements, particularly the inclusion of Prime Video, transformed Prime into a multifaceted service. By offering access to streaming content alongside expedited delivery, Amazon significantly broadened its appeal. This strategic evolution marked a turning point, and by the 2010s, Prime's enhanced value proposition had attracted millions of new subscribers, establishing it as one of Amazon's most recognizable and profitable offerings. Today, Prime serves as a cornerstone of Amazon's success, exemplifying its ability to adapt to consumer needs and leverage innovation to sustain growth.

Between 2013 and 2021, Amazon Prime experienced remarkable growth, with its net sales increasing by approximately $29 billion and reaching a total of $31.77 billion by 2021 (Search Logistics, 2022). This surge was largely fueled by the COVID-19 pandemic, which reshaped consumer habits and drove significant growth in e-commerce. During this period, from 2019 to 2021, Prime added approximately 28 million new subscribers as individuals increasingly relied on online platforms for essential goods and entertainment.

As of 2025 Amazon Prime has **just over 200 million** members worldwide (increv.co/academy/ amazon-prime-users). Analysts project that membership could rise to approximately **300 million** by 2030 (capitaloneshopping.com/research/amazon-prime-statistics).

Amazon Prime has fundamentally reshaped how people shop, and its influence extends far beyond the subscription fee itself. By bundling fast shipping with exclusive deals and services like Prime Video, Amazon pulls customers deeper into its ecosystem and gives them fewer reasons to shop elsewhere. The result is not just loyalty, but dependency.

Prime members consistently spend far more than non-members. Recent U.S. data shows the average Prime customer spends roughly $1,170 per year, compared to about $570 for non-Prime shoppers, nearly double the annual spend (Red Stag Fulfillment, 2024). That gap holds across spending tiers as well. About 39 percent of Prime members spend more than $200 in a given period, while only 13 percent of non-Prime customers reach the same level (Business Dasher, 2024).

These numbers underscore Prime's real value to Amazon. It is not just a membership program, but a powerful behavioral engine that increases purchasing frequency, raises spending ceilings, and locks customers into Amazon's broader platform.

Prime's real power was not the annual fee. It was the behavior it created. Once customers trusted Amazon to be fast, convenient, and reliable, they stopped comparing every purchase and started defaulting to the platform. That loyalty became fuel for the entire network. More orders meant more volume. More volume meant more pressure. And that pressure eventually moved from the customer promise into the buildings, trucks, routes, and people responsible for keeping that promise alive.

Although Prime's initial rollout had a slow start, its steady expansion and the growing acceptance of online shopping have firmly established it as a cornerstone of Amazon's business model. Over time, Prime has evolved far beyond a simple delivery service to become a comprehensive ecosystem. It now encompasses streaming services, exclusive discounts, and seamless shopping experiences, all of which play a crucial role in enhancing customer satisfaction and loyalty. This multifaceted approach has not only driven Amazon's growth but also reinforced its ability to retain a vast and loyal customer base, making Prime an integral part of its long-term success.

Amazon's operations are guided by four core principles that form the backbone of its identity and enduring success: customer obsession rather than competitor focus, a relentless passion for innovation, an unwavering commitment to operational excellence, and a forward-looking approach grounded in long-term thinking. These principles are more than abstract ideals; they are integral to Amazon's culture and influence decisions, processes, and interactions at every level of the organization.

At the center of Amazon's culture is the idea of Day 1, Jeff Bezos' belief that the company should keep the urgency, speed, and hunger of a startup no matter how large it becomes. From the outside, Day 1 sounds like a call to stay innovative and customer-focused. Inside the operation, it felt more complicated. Day 1 created momentum, but it also created permanent urgency. The work was never finished. The bar was never settled. Every improvement became the new baseline, and every baseline eventually became another expectation pushed down through the network.

"Day 1" embodies Amazon's determination to approach each day with the creativity and drive that fueled its early days, ensuring the company remains adaptive and nimble in a highly competitive environment. This philosophy has enabled Amazon to stay ahead of market trends, consistently reinventing its processes, technologies, and services to deliver exceptional value. It also reinforces a culture of innovation, where experimentation and customer obsession are central to every initiative.

Above all, "Day 1" prioritizes delighting customers, which Amazon considers the cornerstone of its growth and longevity. This relentless customer focus drives the company to exceed expectations, not just by optimizing existing services but by setting new benchmarks in e-commerce, logistics, and technology. For Amazon, it's not merely about competing but about redefining industry standards and continually raising the bar for what customers can expect.

By adhering to these principles, Amazon built a business model that favors long-term positioning over short-term wins, allowing innovation to compound and relevance to be sustained. The Day 1 mindset reinforces constant urgency, encouraging employees to think big, move fast, and treat stagnation as failure. Inside the company, it is not framed as a slogan but as an expectation. Day 1 becomes the justification for relentless pace and unyielding standards, shaping how decisions are made and how success is defined.

Amazon's transformation from startup to global powerhouse was therefore not just a story of scale, innovation, and market dominance. It was also the blueprint for how pressure would eventually be pushed downward through the organization. Every promise made to customers tightened expectations inside the network. Every efficiency gained upstream compressed margins on the floor. The systems that fueled Amazon's rise did not stop at strategy decks or executive meetings. They reached into warehouses, delivery vans, and break rooms, shaping how work was measured, how mistakes were punished, and how people were treated when numbers mattered more than context. What appeared as growth from the outside felt very different inside the operation, where ambition collided daily with human limits.

The Language of the Machine

I found my way to Amazon the same way a lot of people do, through a seasonal job posting looking for extra help ahead of the holiday rush. By that point, I had already built two tech companies and spent years working with startups. Still, I was curious. I wanted to see how Amazon actually operated from the inside, to understand what really made its logistics engine run.

It felt like a deliberate step away from my usual world of founders, pitch decks, and early-stage chaos. I applied for an Associate role at a nearby Distribution Center, not because I needed the job, but because I wanted the experience. I figured it would give me a ground-level view of a system most people only ever encounter through a package arriving at their door.

I had heard all the stories about Amazon being a demanding place to work, and it did not take long to see that those stories were right. The moment I stepped onto the warehouse floor, I could feel the intensity. The entire operation moved at a relentless pace, with efficiency at the center of everything. It was overwhelming, fascinating, and exhausting all at once. Walking into the Distribution Center for the first time was like stepping into a completely different world. The size of the place was staggering, with row after row of towering racks filled with packages. Everything moved in perfect synchronization, with conveyors running nonstop and workers scanning, sorting, and loading at a speed that seemed impossible to maintain. It was easy to see how Amazon had built its reputation for speed and efficiency, but beneath that, it was clear that the system was built to push workers as hard as possible.

My first role was as a warehouse associate, handling picking, packing, and shipping. The job was physically exhausting, with almost no time to pause. Amazon's training was quick and to the point, making sure every worker knew exactly what was expected. The pressure to hit performance quotas was constant, and the expectation was to keep moving at the same intensity from the moment you clocked in until the shift ended. There was no room for slowing down.

One thing Amazon drilled into every worker was its obsession with customer satisfaction. Everything revolved around getting orders out as quickly and accurately as possible, and there was almost no tolerance for mistakes. Whether you were scanning an item, moving a package down the line, or printing a shipping label, everything was tracked. The entire operation was built for maximum efficiency, from how the shelves were organized to how performance was measured.

Beyond the physical grind, working at Amazon was a crash course in problem-solving, teamwork, and adapting on the fly. There was no easing into things. Every shift was fast-paced, unpredictable, and physically demanding. But there was also a sense of camaraderie among workers, a shared understanding that everyone was in it together. The work itself might have been exhausting, but knowing that everyone else was dealing with the same challenges made it a little easier to push through.

That being said, there were plenty of things about the job that drove me nuts. The physical pace was only part of the adjustment. The other part was learning the language. Amazon's daily operations were filled with acronyms, shorthand, and coded phrases that could make the workplace feel like a language test before the work even began. A new associate might walk into the building expecting to learn a job, only to discover that learning the work also meant learning an entire vocabulary. Every process had a label. Every failure had a category. Every metric had an abbreviation. Every conversation seemed to move faster because the people speaking already knew the code.

A typical message from leadership could sound like this: "CX escalated TBA marked as BC, and the notes on the delivery device were not followed by the DSP's DA, resulting in an RTS."

In plain English, that meant a customer complained because a package was not delivered as instructed and had been sent back to the station. But instead of saying that directly, the message was wrapped in layers of shorthand. To someone new, it sounded less like communication and more like a puzzle. You could understand the work and still feel lost in the language used to describe it.

At first, the shorthand seemed practical. Acronyms saved time. Standard terms allowed leaders across different buildings to discuss the same problems without starting from scratch. In a network operating at Amazon's scale, common language mattered. It helped organize complexity. It gave structure to chaos. It allowed one station to speak the same operational language as another station hundreds of miles away.

But over time, that language became more than a communication tool. It became a filter for credibility. The people who knew the acronyms sounded informed. The people who did not could appear behind, even when they understood the work itself. A manager who could speak fluently in metrics, exceptions, and process terms might be heard more quickly than someone trying to explain the human reality behind the numbers. The language became a kind of access badge. Without it, you could be in the room without fully being part of the conversation.

That created pressure for employees and managers to learn not only what was happening, but how to describe what was happening in the approved terms. A late route was not just a late route. A missed scan was not just a missed scan. A staffing shortage was not just a staffing shortage. Each issue had to be translated into the company's operating language before it could be understood, escalated, defended, or corrected. The words became part of the work.

Mastering that language was not about sounding fluent. It was about surviving inside a system where knowing the right words could matter as much as understanding the actual problem. The danger was that language could start to replace reality. Once a problem was reduced to an acronym, it could feel smaller than it was. Once a human struggle became a metric, it could be discussed without emotion. Once a failure was placed into the correct category, the system could move on, even if the people closest to the work knew the real issue had not been fixed.

That was how Amazon's operational language could both clarify and conceal. It helped the company move fast, but it could also distance leaders from the lived experience of the people doing the work. It created precision, but sometimes at the cost of context. It allowed problems to be tracked, but not always understood.

Inside Amazon, learning the acronyms meant learning how the system thought. It meant understanding what the company valued, what it measured, what it ignored, and what it punished. The language was never just a set of terms. It was a map of power.

The Numbers That Run the Operation

Amazon's obsession with performance metrics forms the backbone of how success is measured and accountability is enforced across its logistics network. These numbers are not passive observations. They influence priorities, shape behavior, and determine who must explain a failure when the operation falls short.

Nearly every part of the delivery process is tracked, measured, and compared against a target. Managers monitor delivery outcomes, dispatch times, route completion, staffing, package defects, customer complaints, and dozens of other indicators through dashboards and scorecards. The systems present this information with an appearance of precision and objectivity, but on the floor, those same numbers often translate into urgency, pressure, and difficult tradeoffs for the people expected to deliver the results.

The metrics serve a legitimate purpose. Amazon could not coordinate an operation of its size without measuring performance, and leaders need to know whether packages are arriving on time, whether routes are leaving the station as scheduled, whether workloads are balanced, and whether customers are receiving what they ordered. The problem begins when measurement stops being a source of information and becomes the primary definition of success.

Inside the operation, the numbers were impossible to ignore. They shaped shift expectations, staffing decisions, performance reviews, and leadership discussions. Managers were expected to know where every important measure stood and to act immediately when one began moving in the wrong direction. Falling below a target did not simply mean the operation had encountered a problem. It meant someone would have to explain why.

Delivery performance was divided into several layers. The company measured whether packages reached customers on the first attempt, whether they arrived within the promised time, whether drivers completed their assigned routes, and whether expedited orders met their accelerated delivery commitments. Each measure reflected something customers reasonably cared about. Together, however, they created an environment in which almost every delivery outcome became a number that could be traced back to a driver, a delivery company, a station team, or a manager.

A single unsuccessful delivery could have many explanations. A customer might not be home, a business might close before the driver arrived, or an apartment building might require an access code that was missing or incorrect. A package might be loaded onto the wrong van, while traffic, weather, construction, or a mechanical problem could disrupt the route. Yet once the result entered the reporting system, much of that context disappeared. What remained was a failed delivery attempt and a number moving in the wrong direction.

Dispatch timing carried the same weight. The company measured whether routes left the station on schedule, how many departed late, and how long each delay lasted. These measures mattered because a late departure could affect the entire delivery day. A driver who left the station behind schedule had less time to complete the route, increasing the likelihood of missed businesses, late deliveries, and packages returning to the station.

The operational logic was straightforward: if a delay could be measured, it could be reduced, and if it could be reduced, it could eventually be eliminated. Every minute became a potential source of improvement. That thinking produced remarkable discipline, but it also turned time into another form of pressure.

Station teams learned to prepare packages, stage carts, position vans, and move drivers through load-out with extraordinary speed. At the same time, delays caused by factors outside the team's control still became part of the station's performance record. Late-arriving volume, missing equipment, staffing shortages, routing errors, and system interruptions all had to be absorbed without allowing the customer promise to slip.

Route planning created another layer of measurement. Amazon tracked the number of routes assigned to each delivery company, the number of packages placed on each route, the number of carts a driver had to load, and the expected workload across the day. The purpose was to distribute volume efficiently and prevent one part of the operation from becoming overloaded. On a dashboard, the routes could appear balanced, but on the road, they were rarely equal.

A route with fewer packages could still be difficult because of apartment complexes, gated communities, commercial stops, rural distances, parking restrictions, or heavy traffic. Another route with more packages might move quickly because the stops were close together and easy to access. The system could count packages and estimate time, but it could not always capture what the driver would physically encounter.

That gap between the planned route and the lived route was where much of the pressure accumulated. The system expected the driver to meet the plan, even when reality made the plan unrealistic. When the driver fell behind, the performance number still recorded the shortfall without fully capturing the conditions that produced it.

Staging and loading were measured with the same intensity. The company tracked when package preparation began, whether carts were ready on time, how quickly drivers completed loading, and how long each wave of vans occupied the loading area. These were the final steps before the packages left the building, and even a small delay could ripple through the rest of the day.

Managers were expected to intervene immediately when the flow began to slow. They moved associates, reorganized staging areas, redirected carts, corrected package errors, and worked with delivery companies to keep vans moving. There was little time for extended analysis while the operation was unfolding because the immediate priority was recovery. This was one of the defining characteristics of Amazon's management system: the metrics did not simply record what had happened; they demanded action while it was happening.

Quality was measured just as aggressively. The company tracked package handling errors, scanning defects, missorts, and customer reports of packages that were shown as delivered but could not be located. Each issue represented a breakdown in the delivery promise, and each could trigger an investigation. Missing-package reports received particular attention because they placed the company's records in direct conflict with the customer's experience.

The system might show that a package had been delivered while the customer insisted it was not there. The cause could be theft, an incorrect delivery location, a misleading scan, or a package left somewhere the customer could not find. Whatever the explanation, the event required a response. Delivery companies and station managers were expected to investigate, document what happened, and identify steps that would prevent it from happening again.

Ambiguity was rarely treated as an acceptable conclusion. The system expected a cause, an owner, and a corrective action, even when the evidence did not support a clean answer. That expectation made the process appear decisive, but it could also force complicated events into explanations that were more certain than the facts allowed.

When an important performance measure fell below target, the company initiated a formal failure review. The review was supposed to examine the breakdown from end to end, identify contributing factors, document root causes, assign corrective actions, and establish deadlines. In theory, the purpose was prevention. If the organization understood why a failure occurred, it could correct the process and keep the same problem from returning.

In practice, the process often felt less like problem-solving and more like a formalized search for accountability. During my time at Amazon, I authored thousands of these reviews, and many required me to explain outcomes that were well outside my direct control. A driver arriving late, an unexpected spike in package volume, an incomplete route, a traffic accident, or a system problem could pull an important measure below target. Once that happened, the expectation was clear: document it, explain it, and own it.

The review process demanded more than data analysis. It required diplomacy because drivers were trying to finish difficult routes, delivery company owners were balancing staffing and financial pressure, and station teams were attempting to move packages through a tightly timed process. Senior leaders wanted a clear explanation and a plan that promised the failure would not happen again. Finding the root cause often became a negotiation over whose version of the event would become the official record.

One group might blame the route plan, while another might point to a late driver. The driver might blame package volume, traffic, or an unrealistic sequence, while the station might identify a staffing shortage or late-arriving packages. Each explanation could contain part of the truth, but the review process pushed toward a single cause and a clearly assigned owner. Reality was rarely that clean.

Weather, traffic, volume surges, missing information, system interruptions, and human fatigue routinely disrupted the operation. These were not unusual exceptions but normal parts of daily logistics. Yet the performance process often treated them as obstacles that should have been anticipated and absorbed. The expectation was not simply to manage uncertainty but to prevent uncertainty from appearing in the results.

The central performance dashboard reinforced this system. It consolidated operational measures into one place, allowing leaders to monitor delivery outcomes, dispatch performance, package defects, customer complaints, and station productivity. When a measure dropped below an acceptable threshold, the dashboard required a written explanation and a corrective-action plan.

This created visibility across the organization. Leaders could see where performance was slipping and compare one station, region, or delivery company against another. Used thoughtfully, that visibility could help identify patterns and direct resources toward genuine problems. But the dashboard also turned explanation into a requirement that could be measured on its own.

The system did not only track whether the performance target had been missed. It also tracked whether the required review had been completed. That distinction became increasingly important because I rarely received meaningful feedback on the formal failure reviews I submitted. I could write a detailed analysis, identify a recurring operational problem, explain the contributing factors, and recommend a solution, yet receive no acknowledgment that anyone had reviewed the substance of what I wrote.

When a review was missing, however, the response was immediate. That contrast made the organization's priority clear because the existence of the review appeared to matter more than the quality of the analysis inside it. The system could confirm that the requirement had been completed, but it could not confirm that anyone had learned from it.

Eventually, I began to question whether many of the reviews were being read at all. On several occasions, I entered text that did not constitute a meaningful operational analysis, yet no one questioned it, asked for clarification, or appeared to notice. The result was difficult to ignore. When a review was absent, the system detected the omission immediately, but when something was entered into the required field, its substance seemed almost irrelevant.

That experience changed how I understood the process. The problem was larger than inattentive leadership because the structure itself rewarded documentation more reliably than judgment. The organization could prove that a failure had been reviewed without proving that the review produced understanding, action, or improvement.

This did not mean that metrics were useless. Many of them revealed real problems and helped managers identify delays, locate defects, balance workloads, and protect customer commitments. Without measurement, an operation of Amazon's scale would become unmanageable. The danger came from treating the numbers as complete representations of reality.

A metric could show that a route left late, but not fully explain why. It could show that a driver failed to finish, but not capture the conditions on the road. It could show that a package was missing, but not immediately reveal whether the cause was theft, an incorrect scan, a bad address, or a customer misunderstanding. It could show that a station missed a target, but not identify who absorbed the pressure required to recover the operation.

The numbers were not necessarily false, but they were incomplete. Once they appeared on a dashboard, however, they gained authority. The numerical version of the event became more powerful than the human explanation behind it. Employees and managers learned that understanding the problem was not enough because they also had to translate it into the language the reporting system would accept.

That shaped behavior throughout the operation. People focused on the measures most likely to attract attention, prioritized visible failures over less measurable problems, and learned how to explain performance in the approved format. They also learned how to demonstrate that corrective action had been taken, even when the underlying condition remained unresolved. Over time, the process could become less about improving the work and more about satisfying the system that monitored it.

Amazon often described its metrics as objective, fair, and data-driven. Inside the operation, they also functioned as instruments of control because they determined what received attention, who had to defend an outcome, and which version of events entered the official record. The company's data-driven approach was one of the reasons it achieved logistical dominance, but that same system placed relentless demands on the people responsible for producing the numbers.

There was little space to slow down, question the model, or acknowledge that some variation could not be eliminated. Success became synonymous with keeping the numbers within range, while falling behind brought scrutiny, explanations, corrective actions, and consequences. That was the deeper lesson: metrics did not merely describe Amazon's operation; they shaped it. Once the numbers became the primary definition of success, the system did not ask who absorbed the pressure. It asked whether the target was met.

The Systems Behind the Promise

Amazon's internal systems are often described as productivity aids, but inside the operation they function more like an operating system for human behavior. They do more than organize work. They monitor it, guide it, correct it, and escalate it in real time. Each screen, scan, alert, and dashboard becomes part of a larger structure that teaches people how fast to move, what to prioritize, and what consequences follow when the work falls outside the expected pattern.

From the outside, Amazon's delivery promise looks simple. A customer places an order, receives a confirmation, and waits for the package to arrive. Inside the network, that promise becomes a chain of decisions, scans, route assignments, labor plans, capacity forecasts, exception reports, and performance reviews. The work is physical, but the rhythm is digital. Packages move through buildings, vans, yards, carts, belts, staging areas, and neighborhoods, but behind that movement is a software-driven system constantly measuring whether reality is keeping pace with the plan.

That is what makes the operation so powerful and so unforgiving. The systems are designed to remove delay, reduce uncertainty, and force visibility into almost every corner of the process. Every package has a status. Every route has a plan. Every shift has a staffing model. Every manager has a set of numbers to defend. When the system works, it creates speed at a scale few companies can match. When it strains, the pressure lands on the people expected to keep the machine moving.

Amazon's distribution and delivery operations rely on a layered ecosystem of planning tools, performance dashboards, routing systems, labor tracking platforms, training portals, issue management processes, and scanning technology. Each layer has a purpose. One system forecasts volume. Another organizes labor. Another plans routes. Another tracks package flow. Another measures delivery performance. Another records exceptions and corrective actions. Together, they create a network where logistics, data analysis, and real-time intervention happen at the same time.

The result is a workplace where decisions are rarely isolated. A volume forecast affects the labor plan. The labor plan affects staffing. Staffing affects load-out speed. Load-out speed affects driver departure times. Driver departure times affect delivery performance. Delivery performance affects leadership calls, scorecards, and corrective-action reviews. A single missed scan, late truck, staffing gap, or routing issue can ripple across the day and appear later as a number someone has to explain.

Performance tracking is the backbone of this logistical machine. Managers are given visibility into what is happening across the operation, often while the work is still unfolding. Delivery progress, package exceptions, route delays, staffing levels, scan accuracy, and station performance can all be monitored as live signals. The expectation is not simply to observe the work. The expectation is to intervene before a problem becomes visible to the customer.

That expectation changes the role of management. A frontline manager is not only leading people. They are also managing the gap between the system's plan and the building's reality. If a route falls behind, if packages are staged incorrectly, if a driver leaves late, if volume exceeds forecast, or if labor is short, the system makes the deviation visible. Once visible, it demands explanation. Visibility becomes pressure, and pressure becomes action.

Leadership dashboards take this oversight even further. They consolidate key performance indicators into a format that allows performance to be reviewed daily, weekly, and across longer trends. These dashboards can be useful because they help leaders see patterns that might otherwise be missed. A decline in delivery success, a spike in package defects, a change in load-out timing, or a recurring staffing issue can all point to a deeper problem.

But the dashboard also changes the nature of accountability. Once a number drops below expectation, the system often requires a documented response. A leader may be asked to explain what happened, why it happened, what corrective action was taken, and how recurrence will be prevented. In theory, this creates discipline. In practice, it can become a ritual of explanation, where the immediate need is not always to solve the root cause but to produce a response that satisfies the review process.

This is one of the central tensions inside a metrics-driven operation. Measurement can reveal problems, but it can also narrow attention. When a system is built around thresholds, exceptions, and escalations, people learn which numbers matter most. They learn which problems generate scrutiny and which problems remain invisible because they are harder to measure. The operation becomes skilled at responding to what the system sees, even when the system does not see everything.

Routing and dispatch systems are just as critical to last-mile logistics as the software that controls package movement inside the building. Before drivers arrive, packages must be grouped, sequenced, assigned, and prepared for delivery. The routing process determines how the day will feel for the people executing it. A route is not just a line on a map. It is a workload. It is a body moving through apartments, businesses, driveways, lockers, stairs, heat, traffic, parking problems, access codes, and customer expectations.

These routing systems are built to optimize efficiency, but efficiency on a screen does not always capture the full reality of the road. A route may look balanced by package count but become difficult because of commercial stops, long travel distances, apartment complexes, traffic patterns, late arrivals, missing access information, or unrealistic sequencing. The system can estimate the work, but the driver lives it.

Dispatch planning systems reduce much of the manual work that once went into organizing delivery operations. They help assign routes, plan staging, and guide the flow of vans through the morning launch. Used well, they can reduce confusion, shorten load-out time, and help delivery teams focus on execution rather than last-minute problem solving. They also make the dispatch process more standardized, which is essential when a network is operating across many sites at high speed.

But standardization has a cost. When the system becomes the primary planner, local judgment can become secondary. Managers and dispatchers may still understand which routes are difficult, which areas create recurring problems, which drivers are new, and which commercial zones require special attention. Yet the system's output often carries more authority than the people closest to the work. The plan arrives with the weight of technology behind it, even when local experience knows the plan may not survive contact with the day.

A key part of last-mile logistics is the ability to adjust when packages do not move according to plan. Misrouted packages, late-arriving volume, incorrect staging, damaged labels, missing scans, and unexpected disruptions are routine parts of the operation. Package-control systems allow managers to reassign, redirect, hold, or correct package movement in real time. This flexibility matters because the network depends on constant recovery. The promise to the customer is built not only on planning, but on the ability to fix what planning missed.

Over time, these systems have become more integrated. Workflows that were once separated across different screens or processes are often consolidated into broader operational platforms. This makes the operation faster and easier to manage in some ways. It also deepens reliance on the system. When one platform becomes the place where packages are tracked, exceptions are corrected, and performance is reviewed, the system is no longer just a tool. It becomes the place where operational reality is defined.

Capacity planning systems add another layer of control. They help leaders anticipate package volume, understand building limits, forecast labor needs, and prepare delivery partners for the expected workload. These tools are essential because the operation cannot afford to guess. Too few people, too little equipment, or too little delivery capacity can slow the entire day. Too much labor or unused capacity creates unnecessary cost. The business depends on balancing demand with resources as tightly as possible.

This balance is one of the reasons Amazon operates with such precision. Volume planning, labor planning, equipment planning, and delivery capacity are tied together. The goal is to have the right number of people, the right number of routes, the right number of carts, the right staging plan, and the right delivery coverage at the right time. When that alignment works, the building feels controlled. When it fails, the failure is immediate and visible.

Labor management systems are equally important because they turn human time into operational data. Associates are assigned roles, coded to tasks, tracked through scans, and measured against expected activity. Managers monitor whether labor is being used correctly and whether the recorded work matches the actual work. This is necessary in a high-volume operation because staffing decisions depend on accurate labor data. If the data is wrong, the plan can be wrong.

But labor tracking also changes the worker's relationship to time. Time becomes something that must be explained, coded, reconciled, and justified. A missed scan can make work disappear from the system even if the person was physically working. A miscoded task can distort productivity. A gap in the data can create the appearance of idleness or inefficiency. The system does not always know what happened. It knows what was recorded.

That distinction matters. A worker may have helped clear a jam, moved packages from a problem area, assisted a new associate, searched for a missing package, or responded to a manager's direction. If the action is not captured correctly, the system may not recognize the value of the work. In a workplace governed by dashboards, unrecorded work can become invisible work.

Managers are expected to close those gaps. They review labor records, correct missed entries, reconcile task assignments, and make sure the system reflects the day accurately. This creates another layer of administrative pressure on top of the physical operation. The manager is responsible not only for getting the work done, but for making sure the digital record proves the work was done in the right way.

When operational problems arise, issue-management systems provide a structured way to document and track them. A broken scanner, system outage, delayed truck, routing problem, missing equipment, or recurring defect can be logged, assigned, investigated, and closed. In a network as large as Amazon's, this structure is necessary. Without a formal process, problems would disappear into conversation, memory, or local workaround.

The strength of issue tracking is that it creates a record. The weakness is that a record is not the same as resolution. A problem can be documented without being truly fixed. A ticket can be closed while the underlying condition remains. The system can show that action was taken, even when the frontline still experiences the same obstacle the next day. Documentation becomes part of the work, but it does not always replace the need for local judgment, persistence, and follow-through.

Training systems serve another function. They standardize what employees are taught, when they are taught it, and how completion is recorded. This allows the company to train large numbers of people across different roles and locations while maintaining consistency. Employees can be assigned role-specific modules, safety instruction, compliance training, and operational procedures. For a company operating at Amazon's scale, this kind of structure is unavoidable.

Yet training inside a high-pressure operation is often shaped by time. The module may be completed, the record may be updated, and the requirement may be satisfied, but real understanding usually comes from doing the work under pressure. A person can learn the policy on a screen, but the building teaches them what matters. It teaches them which metrics create urgency, which mistakes get escalated, which shortcuts are tolerated, and which problems must be solved quietly before they become visible.

Sortation and scanning systems form the physical-data connection at the center of the operation. Every scan tells the network where a package is, where it should go, and whether it is moving correctly. Scanning creates order out of volume. It allows packages to be sorted into containers, routed to delivery areas, staged for drivers, and tracked through the final mile. Without accurate scanning, the operation loses visibility. Without visibility, the system loses control.

This is why scan discipline becomes so important. A package in the wrong location is not just misplaced. It becomes a defect. A missed scan is not just a small oversight. It becomes a data gap. A wrong container, wrong route, wrong staging location, or wrong status can create delays that surface later in the day. The physical mistake becomes a digital exception, and the digital exception becomes a performance problem.

Together, these systems form the foundation of Amazon's operational framework. They maintain efficiency, reduce disruption, and support the speed customers have come to expect. They allow the company to coordinate massive amounts of work across buildings, drivers, delivery partners, managers, and support teams. They are impressive in scope, and their effectiveness is part of why Amazon became one of the most powerful logistics companies in the world.

But these systems also create the pressure-cooker environment that defines the operation. Everything is measured. Everything is tracked. Everything can be reviewed. Workers and managers are expected to keep pace with a system that never slows down and rarely accepts the messiness of human limits as a valid explanation. The machine does not feel the heat in the yard, the pain in someone's back, the confusion of a new hire, the fatigue of a driver, or the judgment call a manager makes in the middle of a chaotic morning. It records signals, compares them to expectations, and demands correction.

Mastery of these systems is often framed as empowerment. In some ways, it is. A manager who understands the tools can see problems earlier, move resources faster, and protect the operation from failure. An associate who understands the process can work more effectively and avoid mistakes. A dispatcher who understands the routing logic can make better decisions before the vans leave the building.

But mastery is also a prerequisite for survival. The systems do not replace judgment, but they constrain it. They shape what managers notice, what employees prioritize, and what leaders reward. They determine which problems become visible and which remain hidden. They create the language of accountability and the boundaries of acceptable performance.

Understanding these systems is essential because they are not just technical infrastructure. They are the mechanisms through which Amazon's values are enforced at scale. Speed becomes a dashboard. Efficiency becomes a labor plan. Customer promise becomes route pressure. Accountability becomes a corrective-action record. Human effort becomes data.

And once work becomes data, the system becomes more than a tool. It becomes the boss.

Amazon Leadership Principles

At Amazon, the culture and constant push for innovation all trace back to its Leadership Principles. They're not just corporate catchphrases. They guide how people think, how they make decisions, and how they show up to work each day. **Customer Obsession** is the one that stands out the most. It puts the customer at the center of everything, and you can see that clearly in how Amazon Prime evolved. What began as a simple free-shipping perk turned into a full subscription service with Prime Video, Prime Music, same-day delivery, and a long list of other benefits. None of that was added just to pad the offering. It all came from Amazon's drive to make things easier and more convenient for the customer.

The "no-questions-asked" return policy is another good example. It made returning items effortless, built a huge amount of trust, and pushed competitors to tighten up their own standards just to keep up. In recent years, though, Amazon has started to pull back a bit. Too many people abused the system, and the company responded with stricter return rules in certain cases.

Ownership is another principle that defines Amazon's culture. It encourages employees to take full accountability for their work, as though they own the company themselves. That mindset has driven some bold, long-term decisions, including the creation of Amazon Web Services (AWS). At a time when cloud computing was barely on most companies' radars, Amazon invested heavily in building its own infrastructure, which now powers a significant portion of the internet. The success of AWS shows what can happen when leaders focus beyond immediate results and are willing to think long term. That same mindset filters down through the organization, pushing employees to approach problems with a "how can I fix this?" attitude instead of passing responsibility to someone else.

The principle **Invent and Simplify** captures Amazon's instinct for innovation and its drive to find straightforward solutions to complex problems. Amazon Go stores are a clear example, using sensor technology to eliminate checkout lines entirely. Walking out of a store without stopping at a register might have sounded like science fiction not long ago, but Amazon made it real. That same thinking is behind products like the Kindle and Alexa. The Kindle did not just digitize books. It reshaped how people read. Alexa brought voice commands into everyday routines. These innovations are not about flashy technology. They are about making life simpler and more convenient for customers.

Are Right, A Lot reflects Amazon's belief in good judgment, bolstered by a willingness to test ideas rigorously. Alexa's development is a great example of this principle in action. It didn't become the smart assistant we know overnight. It took countless iterations, feedback from users, and testing in thousands of different scenarios to improve its ability to understand natural language. The same approach underpins Amazon's recommendation algorithms, which are constantly fine-tuned to make them more relevant and helpful. This kind of careful, data-driven refinement ensures that Amazon keeps hitting the mark.

Learn and Be Curious highlights Amazon's emphasis on continual learning and growth. The Career Choice program, which covers up to 95% of tuition for employees pursuing high-demand skills, is a standout example of how the company supports personal development. Beyond that, teams at Amazon are encouraged to attend industry conferences and bring back fresh ideas. It's this kind of openness to new knowledge that helps Amazon stay ahead of the curve in a competitive landscape.

Another guiding principle, **Hire and Develop the Best**, underscores Amazon's focus on talent. Every hiring decision is treated as an opportunity to raise the bar, which is why the interview process is famously rigorous. Once onboard, employees are supported through mentorship and leadership programs designed to help them grow. The company does not just hire smart people. It invests in them, making sure high-potential employees have the tools and opportunities to develop into key roles.

Insist on the Highest Standards is one of those principles that really defines Amazon's customer-facing reputation. For example, the company's packaging is meticulously designed to minimize damage during shipping, reducing returns and enhancing customer satisfaction. In customer service, representatives are trained to solve problems thoroughly, aiming for real solutions rather than quick fixes. It's this attention to quality that sets Amazon apart in the eyes of its customers.

Think Big is a principle that has shaped many of Amazon's boldest moves. When the company ventured into streaming with Prime Video, it did not stop there. The launch of Amazon Studios and the push into original content were not simply about competing with Netflix. They were about building something new for its audience. That kind of ambitious thinking has helped Amazon carve out new markets repeatedly and expand far beyond its original boundaries.

Bias for Action reflects Amazon's sense of urgency and its willingness to take calculated risks. The rapid expansion of fulfillment centers and delivery hubs, which allowed the company to meet soaring demand for fast delivery, is a clear example. Amazon's early investment in autonomous drone technology through its Prime Air project reflects the same principle in practice. It shows a readiness to experiment with emerging technology in order to stay ahead.

Frugality is about doing more with less. In its early days, Amazon operated under tight financial constraints, and that mindset never disappeared. The use of robots in fulfillment centers today is not simply about cutting costs. It reflects an effort to scale operations in smarter and more efficient ways.

At Amazon, frugality is not about being cheap. It is about being resourceful, disciplined, and creative.

Earn Trust reflects Amazon's belief in transparency and humility. Jeff Bezos himself set the tone by openly acknowledging failures like the Fire Phone, showing that the company isn't afraid to admit when it gets things wrong. This openness builds trust, both internally and externally, and creates a culture where improvement is always possible.

Dive Deep emphasizes the importance of understanding the nitty-gritty details. Leaders are expected to roll up their sleeves and stay connected to what's happening on the ground. Whether it's reviewing operations in a fulfillment center or digging into customer feedback, this hands-on approach ensures that decisions are well-informed.

Have Backbone; Disagree and Commit encourages healthy debate. Teams at Amazon are expected to challenge each other's ideas during the decision-making process. A perfect example of this is the development of the Echo, which involved heated discussions about the feasibility of voice AI. Once the decision was made, though, everyone got behind it, resulting in one of Amazon's most successful products.

Amazon's Leadership Principles are presented as universal truths, but in practice they function as flexible tools, emphasized or ignored depending on who holds power and what outcome is desired. When applied consistently, they can encourage ownership, rigor, and accountability. When applied selectively, they become a shield for poor leadership and a weapon against those who challenge broken processes. What matters is not the principles themselves, but who gets to interpret them and when. That gap between stated values and lived reality is where much of Amazon's culture takes shape, and it is where the stories that follow begin.

From Click to Doorstep

From the outside, Amazon's logistics network appears seamless, a finely tuned system moving products from seller to customer with astonishing speed. Inside the operation, that same journey is a chain of handoffs, each one carrying risk, pressure, and consequence for the people responsible for keeping packages moving. Every scan, transfer, and deadline tighten expectations downstream. What looks like efficiency at scale depends on thousands of small decisions being made correctly, often under time constraints that leave little room for recovery when something goes wrong.

The journey begins with the First Mile, where products enter Amazon's vast logistical ecosystem. This stage starts when items are sourced from sellers, manufacturers, and vendors before being funneled into Amazon's Fulfillment Centers. These facilities serve as the backbone of inventory management, where an intricate system of high-tech automation and carefully designed storage strategies ensures that products can be retrieved quickly and efficiently.

Once products arrive at the fulfillment center (FC), they are scanned, sorted, and stored using algorithms that prioritize accessibility. When an order comes in, the system determines the fastest and most efficient way to retrieve, pack, and ship the item. Sellers rely on Amazon's Seller Central platform to monitor and manage their inventory, while Amazon's retail teams coordinate directly with manufacturers to maintain a selection that meets the demands of its vast customer base. Behind the scenes, Amazon's freight network and third-party carriers ensure that inventory continues to flow, keeping products stocked and ready for immediate fulfillment.

Amazon doesn't store every product it offers in its fulfillment centers. That would be impossible. When an item isn't stocked at an FC, the vendor ships it directly to the customer. If that vendor is based in the United States, the package usually goes through USPS, though UPS and FedEx are also common depending on where the customer lives. Outside the U.S., Amazon often relies on each country's local postal service when its own delivery network can't cover the area.

Every process within the First Mile is built to eliminate inefficiencies, ensuring that when a customer places an order, the product is ready to ship without delay. This stage is not just about storage and stock management. It is a carefully synchronized operation that sets the foundation for Amazon's ability to deliver at an unprecedented scale.

Once an order is packed and ready for transit, the Middle Mile takes over, bridging the gap between fulfillment and delivery. This stage involves the large-scale movement of packages between Fulfillment Centers (FC), Sortation Centers (SC), and Distribution Centers (DC). To handle this volume, Amazon leverages its dedicated fleet of trucks and its growing air cargo network, Amazon Air, to keep shipments moving quickly and efficiently.

This phase is all about precision. Every package needs to arrive at the right place at the right time, and that level of coordination requires a tightly controlled system. Sortation Centers play a key role in this process, organizing packages by their final destinations before they are handed off to the next stage. Behind the scenes, advanced technology powered by machine learning continuously refines delivery routes, optimizing schedules to save time and control costs. These efforts become even more critical during peak shopping seasons, when the sheer volume of packages could overwhelm a less sophisticated system.

The Middle Mile is really the backbone of Amazon's delivery network. It runs quietly in the background, but it carries a huge amount of the workload. It never gets the same attention as last-mile delivery, yet this is where most of the heavy lifting happens. Whether it's keeping up with Prime's next-day promise or making sure something ordered at two in the morning is already on the move by sunrise, the Middle Mile is what keeps the whole machine from falling behind.

At the FC, packages are sorted by the delivery regions that location supports. Once sorted, they are loaded onto Amazon trucks either on pallets, in Gaylords, which are large cardboard boxes that sit on a pallet, or as fluid loads, where loose boxes are stacked directly into the trailer. Fluid loads are a nightmare for associates to unload at the SC, but they still occur. Those trucks then head to the SC, where everything is sorted again and reloaded into trailers bound for the Distribution Center (DC), marking the start of the Final Mile.

When packages arrive at the DC, they're sorted again, this time by ZIP code, so they can be sent out to customers. The Final Mile is the part of Amazon's logistics system people actually see, and it's also where things get the most complicated. This is the last stretch of the journey, the moment when a package reaches someone's doorstep, but getting it there is anything but simple. Drivers deal with unpredictable road conditions, traffic that can throw off an entire route, and customers who decide to update instructions at the last second. Some stops involve apartment buildings with confusing entry codes. Others mean dodging aggressive dogs or trying to figure out where a customer wants the package dropped when the notes barely make sense.

To handle all these challenges, Amazon built a layered delivery system, and at the center of it is the Delivery Service Provider program, or DSPs. These are small business owners who run fleets of Amazon-branded vans and cover local routes in cities and towns across the country and around the world. If you've ever seen those dark gray-blue vans flying through your neighborhood, that's the DSP network at work. They take on most of Amazon's last-mile deliveries, giving the company the consistency and scale, it needs to keep packages moving without missing a beat.

Even with DSPs handling a massive portion of the workload, Amazon still needs backup during high-demand periods like the holiday rush. That is where Amazon Flex comes in. Flex drivers are independent contractors who use their personal vehicles to deliver packages, following instructions from the Amazon Flex app. This model allows Amazon to rapidly scale its delivery capacity when demand spikes, ensuring that customers still get their orders on time. It is an efficient and cost-effective solution for Amazon, but it also raises ongoing debates about worker classification, job security, and fair compensation. As the reliance on gig workers continues to grow, so do the questions about how these workers fit into Amazon's long-term logistics strategy.

The Final Mile is the true test of Amazon's logistics network. It is the part of the process that customers see and experience firsthand, which leaves almost no margin for error. Whether it is a DSP driver unloading packages from a branded van or a Flex driver racing to meet a tight schedule, this stage is where Amazon's ability to navigate real-world challenges is fully exposed. Every delivery becomes an opportunity to reinforce the company's promise of speed, reliability, and efficiency, or to fall short of it.

Innovation sits at the center of Amazon's Final Mile operations, and the company is constantly looking for ways to refine and reinvent how deliveries are made. One example is Key In-Garage Delivery, a Prime-exclusive service that allows drivers to place packages and groceries directly inside a customer's garage. It addresses several problems at once, including package theft, weather damage, and the frustration of not being home when a delivery arrives. For customers who want an alternative to doorstep drop-offs, Amazon Hub lockers offer another option. Found in apartment buildings, grocery stores, and shopping centers, they give customers the flexibility to pick up orders on their own schedule.

In 2018, Amazon partnered with GM's OnStar and Volvo On Call to launch In-Car Delivery in the U.S., a feature that allowed Prime members to have packages placed inside eligible vehicles parked in public areas. I was part of the team in

Portland when this program rolled out, and for the most part it worked surprisingly well, as long as the customer's car was parked reasonably close to the delivery address.

Another major focus of Amazon's Final Mile strategy is giving customers as much visibility as possible into their package's journey. Features like live tracking, estimated arrival times, and photo confirmations remove much of the guesswork and allow customers to stay informed at every step. This level of transparency does more than improve convenience. It gives customers a sense of control over their deliveries and reinforces Amazon's reputation for reliability.

These innovations are not just about improving logistics. They are part of Amazon's broader effort to transform the delivery experience itself. By prioritizing convenience, flexibility, and transparency, Amazon is not simply moving packages but shaping how modern e-commerce integrates into everyday life. The goal is not just to meet expectations but to create a system that customers actively trust and depend on.

What keeps this entire network running is Amazon's deep reliance on technology and data. Real-time analytics, machine learning, and predictive algorithms work together to identify inefficiencies and drive continuous improvement. These systems influence everything from inventory placement in the First Mile to route optimization in the Middle Mile and delivery execution in the Final Mile. This interconnected, data-driven approach ensures that packages move quickly, accurately, and consistently, no matter the volume or external challenges.

Amazon's logistics network is more than a system for transporting goods. It reflects the company's relentless pursuit of efficiency and innovation. By mastering each stage of the journey, Amazon has redefined expectations for e-commerce logistics, raising the bar for speed and reliability across the industry. This highly optimized system, driven by advanced technology and an unwavering focus on customer satisfaction, underpins Amazon's dominance. It is what enables the company to deliver millions of packages on time, every time, and maintain its position as a global leader in logistics and e-commerce.

Amazon's logistics network is not just a transportation system. It is a pressure system. Every stage is designed to remove delay, reduce variation, and protect the customer promise. That design is what makes Amazon so powerful, but it also means that by the time a package reaches the warehouse floor, the margin for error has already been compressed by every decision made before it arrived.

Inside Amazon's Fulfillment Center

A fulfillment center is loud before it is impressive. The customer clicks "Place your order" and sees a clean confirmation screen, but inside Amazon, that click becomes motion almost instantly. Systems locate the inventory, assign priority, trigger fulfillment, and begin moving the order through a network built around speed. What feels effortless from the outside becomes noise, scans, conveyor belts, pressure, and people moving fast enough to keep the promise alive.

Conveyor belts hum without pause. Scanners beep in uneven rhythms. The air carries cardboard dust, machine oil, and recycled air that never quite feels fresh. Associates move quickly through narrow aisles and tightly timed workflows where hesitation is visible and mistakes travel downstream. It is a space built for motion, not comfort, and everything about it signals urgency long before a metric appears on a screen.

As soon as the order is received, the system checks for the nearest Fulfillment Center that has the item in stock. If it is available locally, the fulfillment process begins immediately, with warehouse systems prioritizing the order for packing and shipment. If the item is not in stock at a nearby location, Amazon's system determines the next best option. This could mean shipping the product from another Fulfillment Center across the country or notifying a vendor to send the item directly using external carriers like USPS, UPS, or FedEx. While this entire process happens in a matter of seconds, it represents an intricate decision-making system that blends automation, logistics, and human oversight into one seamless flow.

What makes this fulfillment system remarkable is the balance between speed and precision. Amazon has spent years refining its network to ensure that every step, from identifying available inventory to routing shipments, is executed with near-perfect timing. It is not just a logistics process. It is an intricate fusion of advanced technology and human effort that gives Amazon its competitive edge and allows it to promise, and often deliver, unmatched efficiency.

Once an item is confirmed for fulfillment, the systems at the warehouse spring into action, locating its exact position within the facility. While this step might seem straightforward, it is anything but simple. Inventory is stored in a highly strategic way, depending on factors like size, weight, and demand frequency. Some products are placed on movable shelves that can be retrieved by robotic systems, while others are housed in static storage areas that require human associates to pick them manually. The system determines the most efficient retrieval method, ensuring that the order moves as quickly as possible from storage to packing, keeping the fulfillment machine running at full speed.

If the item is stored on a movable shelf, the robots take over. These compact, square devices appear simple, but their function is anything but. Equipped with retractable tops, they can lift entire shelving units off the ground. The system identifies the closest available robot to the requested shelf and, with seamless precision, sets the retrieval process in motion. What follows is nothing short of mesmerizing. Hundreds of robots move in synchronized patterns across the warehouse floor, resembling a perfectly rehearsed dance.

Each robot operates independently, yet their movements are intricately coordinated, avoiding collisions while efficiently locating via RFID and lifting the correct shelves. Watching them in action feels like observing a live performance, where technology and choreography meet. This intricate dance is a testament to the engineering and innovation that underpins Amazon's ability to fulfill orders quickly and reliably. It's not just about retrieving an item; it's about doing so with an efficiency that pushes the boundaries of what modern logistics can achieve.

The robot glides beneath the designated shelf, its retractable top extending with precision to lift the entire unit a few inches off the ground. Once secure, the robot navigates the bustling warehouse floor to deliver the shelf to a Picker, a human worker who plays a crucial role in this hybrid system. Amazon's AI ensures the robot directs the shelf to the nearest available Picker station, eliminating idle time and maximizing efficiency.

Each movable shelf is a compact, four-sided structure brimming with products, and the system rotates the unit to present the side containing the requested item. The robot even raises the shelf slightly, positioning it at an ergonomic height to minimize physical strain on the Picker. This thoughtful design reflects Amazon's commitment to balancing speed with the well-being of its workers.

Once the shelf is in place, the Picker retrieves the requested item and places it into a container, commonly a box or padded envelope, known internally as a "jiffy." In certain cases, the product's container is used. At this stage, a barcode sticker is printed and automatically affixed to the container. This barcode is pivotal, it transforms the item from warehouse inventory into an individualized shipment, enabling it to be tracked at every subsequent step of the journey. The Picker then transfers the container to a conveyor belt, setting it on its way to the next stage of processing.

If you ever get the chance to tour an FC with robotics, take it. I promise it's an experience you'll never forget. Watching hundreds of robots glide across that massive floor is like watching a strange kind of ballet. They weave around each other with this single-minded drive to complete their routes from start to finish. For good reason, the active floor is completely off limits to anyone on foot. Every now and then, a robot will get out of alignment and tip a shelf, spilling everything onto the concrete. When that happens, the whole grid shuts down so an Area Manager can walk out, rebuild the shelf, and clear the fault. The manager has to wear a special vest that signals the robots to steer clear. Without it, it would be a disaster.

There was a rumor about an FC on the East Coast where an associate ignored the rules, climbed over the safety barrier, and walked straight into the middle of the robot floor. One of the robots latched onto the associate's leg and dragged them across the ground, causing serious injuries. Whether every detail of that story is true or not, the point stands. These warehouses are dangerous places if you don't respect the rules.

This process showcases the intricate collaboration between human workers and robotic systems. While Amazon continues to invest heavily in automation, hinting that robotic advancements may one day assume even the Picker's responsibilities, the current synergy between human intelligence and machine precision creates an extraordinary example of efficiency and innovation. For now, this blend of human touch and technological ingenuity highlights a fascinating moment in the evolution of fulfillment systems, where cutting-edge machines and human adaptability combine to redefine what's possible in logistics.

As the container moves along the conveyor belt, automated scanners read the barcode and log the item's details into Amazon's system. This step enables precise tracking and allows the package to receive its mailing label, which contains critical routing and delivery information. As the package continues downstream, mechanical diverters activated by the barcode redirect it into a designated chute. These chutes funnel packages into specific containers, such as Gaylords, pallets, or carts, based on their final destinations. Gaylords are large six-foot cardboard boxes that may feel like a throwback in a high-tech environment, but they remain a practical and reliable staple in Amazon's warehouse operations for handling bulk shipments.

Once the sorting process is complete, these containers are prepared for transport. Gaylords, pallets, and carts are loaded into Amazon's 53-foot linehaul trailers or box trucks, which form the backbone of the Middle Mile. Amazon uses several loading techniques to maximize efficiency and maintain stability during transit. One common approach is organized palletization, where packages or goods are securely stacked on flat platforms, typically made of wood, plastic, or metal, and arranged to simplify handling and transport. This method helps keep items stable in transit and significantly improves the speed of loading and unloading, reducing the amount of time trailers sit idle at docking stations.

Amazon employs an alternative technique known as fluid loading, where packages are tightly packed into trailers to form interlocking walls. This method is incredibly efficient for maximizing trailer capacity, allowing every inch of space to be utilized. However, what it gains in efficiency, it often loses in ease of handling during the unloading process. Associates tasked with inducting these packages face significant physical challenges. Each item must be scanned as it is unloaded to begin the receiving process and the interlocked walls require meticulous dismantling. This creates a labor-intensive and physically demanding environment, especially during peak seasons when the volume of packages surges.

The strain of this process takes a real toll on workers. Bending, lifting, and maneuvering packages from densely packed trailer walls demand endurance and precision, often pushing employees to their physical limits. While fluid loading supports Amazon's goals of speed and cost efficiency, it also exposes the human cost of those priorities. The company's relentless focus on logistics and throughput frequently places workers in grueling conditions, raising difficult questions about how innovation is balanced against employee well-being. Behind every seamless delivery is a complex system that still relies heavily on human effort, an essential component that too often goes unnoticed.

Once scanned, the packages move onto conveyor belts leading to robotic linear sorters. These advanced machines, guided by barcodes, determine each package's precise destination, directing them into the correct containers for the next leg of their journey. The speed and accuracy of these sorters are remarkable, processing volumes that would be impossible for a human workforce to achieve. This automation is central to Amazon's promise of speed and reliability, keeping the flow of goods uninterrupted as millions of packages make their way through the system each day.

This seamless integration of technology and logistics is the backbone of Amazon's efficiency. Automated systems ensure not only the accuracy of sorting but also rapid dispatch, allowing the network to maintain its relentless pace. However, beneath this marvel of technological precision lies a more complex narrative. While automation has revolutionized operations, it has also displaced countless jobs that once relied on human hands. The relentless pursuit of efficiency and cost savings comes at a human cost, raising difficult questions about the balance between progress and workforce stability. The trade-offs inherent in this innovation underscore a growing challenge: how to navigate a future increasingly dominated by automation while addressing its impact on society.

After sorting, the packages are loaded into linehaul trailers, organized in a way that squeezes out every bit of delivery efficiency. Those trailers head off to their next stops, moving the packages closer to a customer's front door. Every step in this carefully planned chain is another chapter in Amazon's logistics engine, a system that keeps pushing forward with new ideas even as the impacts spill out across the workforce and the larger economy.

Linehaul trailers deliver their loads to Sortation or Distribution Centers, where packages are unloaded, scanned, and verified before moving deeper into the network. Whether loose, palletized, or packed into Gaylords, every package must be accounted for because even a small error can ripple downstream. Yet what keeps a fulfillment center running is not automation alone, but people expected to sustain machine-like speed hour after hour.

Inside Amazon's Sort Centers

Fulfillment centers are built to assemble orders. Sort centers exist to accelerate consequence. This is where volume converges, timelines compress, and small upstream errors become operational problems that can no longer be absorbed quietly. Packages do not pause here. They are redirected, rerouted, and pushed forward at a pace that leaves little room to correct mistakes without disrupting everything downstream.

The Sortation Center sits at the heart of the Middle Mile, playing a critical role in determining the final direction each package will take. In large metropolitan areas, multiple sort centers often operate simultaneously, and it is common for packages to pass through more than one facility before continuing their journey. When a linehaul trailer backs into the dock and the doors open, everything inside is offloaded and fed onto conveyor belts in the intake area. From there, packages are scanned either manually or by robotic scanners lining the belts. In some cases, individual scans are bypassed because the trailer's electronic manifest has already been uploaded, listing every TBA expected inside.

The Track by Amazon number, or TBA, is printed on every package label along with a matching barcode and serves as Amazon's own unique identifier. It links the package directly to the customer's address and always starts with the letters TBA followed by a string of numbers. The code only works inside Amazon's system, but it's the backbone of their entire tracking process. It acts like a digital breadcrumb trail, following the package through every stop in the network. Without it, the whole system would fall apart.

Packages handled by third-party carriers like USPS, FedEx, or UPS work a little differently. Those carriers use their own tracking numbers, completely separate from Amazon's TBA codes. Amazon still displays the tracking information on its website, but the actual number belongs to the carrier delivering it.

One simple tip: if you ever need to contact Amazon customer support about a delivery, giving them the TBA number makes everything go faster. It lets the support team pull up the exact package right away and usually helps resolve issues with far less hassle. It's a small detail, but it shows how Amazon builds its systems around speed, accuracy, and trying to keep the customer experience as smooth as possible.

Once the intake scans are complete, the real sorting begins. Linear sorting machines read each TBA in a split second and push the package down the correct chute onto pallets, Gaylords, carts, or whatever container best fits the volume for the intended Distribution Center. Those containers are then staged and loaded into outbound trailers headed to the DC that will handle the final handoff to drivers. In most SCs, robotics take on a large share of this work, but they rarely handle the full volume. Human labor still plays a critical role in keeping the sort moving. When everything is running clean, it's impressive to watch. But when a package gets stuck or a barcode misreads, the entire flow can back up fast.

Amazon's delivery promise extends even to customers who live in places where Amazon itself doesn't deliver. In those cases, the company relies on USPS to finish the job. Packages bound for those areas are sorted differently, tagged as DDU, or Destination Delivery Unit. That's a USPS term referring to the specific local post office that will make the final delivery. Amazon sorts the package in advance, moves it to the nearest DDU, and hands it off. From there, USPS carriers pick it up and bring it the rest of the way to the customer's door. It's a simple arrangement on paper, but behind the scenes it requires tight coordination to make sure nothing gets lost between systems that operate very differently from each other.

Inside Amazon's Distribution Centers

Distribution Centers are where the network stops being theoretical and starts becoming personal. By the time packages arrive here, the system's tolerance for delay has largely vanished. Every upstream decision, missed scan, and compressed timeline converges at this point before being handed off to the last mile. What happens inside a Distribution Center does not just affect metrics. It determines how much pressure will be transferred to DSPs and, ultimately, to the drivers expected to deliver what the system can no longer absorb.

When packages arrive at a Distribution Center, they enter the final stage of processing before delivery. Linehaul trailers typically arrive between early evening and early morning, carrying the volume that will become the next day's routes. As the trailers are unloaded, each shipment is logged into the station through a pallet or cart scan, an electronic trailer manifest, or, when necessary, individual package scans. These checks confirm that the expected packages reached the correct facility and preserve visibility as they move onto conveyor belts for sorting.

At this point, accuracy is critical. A package sent to the wrong station, placed in the wrong container, or missed during induction can disrupt staging, routing, and driver load-out hours later. Earlier in the network, there may have been time to correct an error. At the Distribution Center, there is almost none left.

Once scanned, each package's TBA number is processed by sophisticated algorithms designed to group packages by geographic zip codes. This systematic sorting process forms the backbone of Amazon's last-mile logistics. By organizing packages with precision, the system ensures they are routed to their final destinations as quickly and efficiently as possible. It's a carefully orchestrated process that underscores Amazon's commitment to operational excellence and its focus on meeting delivery promises.

From there, packages are moved onto conveyor belts, where associates manually scan each item before placing it into designated tote bags. Each tote bag is assigned a specific location, which is scanned again to track its position in the system. This method, often referred to as "sort, pick, and stage," creates a streamlined workflow that sets the stage for precise deliveries.

Overseeing this intricate operation are Area Managers and Shift or Process Assistants, who coordinate with teams of associates to maintain accuracy and efficiency. At this stage, precision isn't just important, it's absolutely critical. Properly sorted packages allow delivery drivers to follow meticulously planned routes, minimizing delays and maximizing both time and resource efficiency. This seamless coordination of human effort and technology is what enables Amazon to uphold its promise of fast and reliable deliveries, even under the immense pressure of high demand.

The next step in Amazon's logistics process is routing, a sophisticated computer-driven operation that analyzes each package and its delivery location to assign it to an optimized delivery route. This step is pivotal, as it lays the groundwork for efficient last-mile delivery. Routes are meticulously planned to minimize travel time and fuel consumption, showcasing Amazon's commitment to logistical precision and environmental consideration.

Once routes are finalized, route totes are loaded onto bakers' racks and staged in designated warehouse areas, ensuring a seamless transition to the delivery phase. At this point, Amazon's Delivery Service Partner (DSP) program and Flex drivers take charge, bridging the final gap between warehouse and doorstep. DSP partners, operating fleets of Amazon-branded vans, manage the bulk of last-mile deliveries with consistency and reliability. Meanwhile, Flex drivers, working as independent contractors with personal vehicles, step in to handle on-demand deliveries, offering the flexibility needed to meet surging demand during peak periods like the holidays.

This adaptable system, powered by advanced routing technology and a diverse workforce, exemplifies Amazon's ability to scale operations efficiently. However, the reliance on contracted partners raises critical questions about labor practices, including worker classification, fair compensation, and job stability. While undeniably effective in achieving operational excellence, this model reflects the duality of progress: groundbreaking efficiency paired with societal concerns. This tension between innovation and its human impact continues to shape Amazon's last-mile logistics, influencing both its operations and the future of work.

Drivers begin their routes by loading pre-sorted totes into their vehicles, adhering to AI-optimized routes tailored for maximum efficiency. These routes, which factor in traffic, weather, and delivery density, streamline the journey and minimize unnecessary travel. At each stop, packages are scanned and photographed upon delivery, providing real-time tracking updates for customers and ensuring transparency. For added security, drivers follow specific instructions to deliver packages to Amazon Lockers, Hubs, or even inside customers' garages, which temporarily grants controlled access to safeguard deliveries against theft or weather exposure. These options exemplify Amazon's commitment to enhancing convenience and addressing customer concerns.

In select regions, Amazon pushes the boundaries of last-mile logistics with its Prime Air initiative, aiming to deliver small packages via drones in under 30 minutes. Though still in its experimental phase, this bold vision for automation faces challenges, including regulatory hurdles, environmental considerations, and questions of practicality. Despite these obstacles, Prime Air highlights Amazon's relentless drive to innovate.

From the click of a button to the final delivery, Amazon's supply chain showcases a seamless blend of cutting-edge technology and human effort. Robotic systems, AI, and AWS infrastructure power each stage, while flexible delivery options and secure solutions reflect the company's focus on customer satisfaction. Yet, this remarkable efficiency comes with significant trade-offs. The relentless physical demands on warehouse workers, the potential displacement of jobs through automation, and the environmental impact of a sprawling logistics network remain pressing concerns.

Amazon's future success depends on its ability to navigate these complexities while balancing operational excellence with social and environmental responsibility. As the company continues to innovate, the challenge lies in ensuring that its progress does not come at the expense of the people, systems, and values that underpin its vision. This balance will define not only Amazon's legacy but also its role in shaping the future of commerce.

Introduction to the DSP Program

Amazon didn't create the Delivery Service Provider (DSP) program out of generosity or to help people "start small businesses," even though that's the message they use publicly. The real reason was scale. By 2017 and 2018, Amazon's package volume had exploded to a level the major carriers could no longer absorb reliably or affordably. Amazon was already expanding its delivery network because package volumes were rising faster than traditional carriers could handle, creating capacity constraints across the industry (Dastin 2017). UPS had begun adding peak surcharges as e-commerce demand overwhelmed its system (United Parcel Service 2017). USPS was facing similar pressure, with its Inspector General warning that the rapid growth in packages was placing increasing stress on the postal network (U.S. Postal Service Office of Inspector General 2017). FedEx soon signaled its own limits as well, eventually pulling back from Amazon volume after determining that the rising package load was no longer profitable to handle (Shepardson 2019).

The DSP program was designed to solve Amazon's last-mile scaling problem by outsourcing the complexity of daily delivery operations. In doing so, it fundamentally shifted operational risk, financial exposure, and accountability away from Amazon and onto a workforce with far fewer protections. DSP owners assumed responsibility for vehicles, staffing, insurance, compliance, and performance, while drivers absorbed the physical strain, safety risks, and metric pressure of an increasingly compressed delivery window. Amazon retained control through contracts, technology, and performance standards, but distanced itself from the direct consequences when the system failed. This arrangement allowed the network to scale rapidly, while insulating the company from many of the human and financial costs created by its own expectations.

All of this pushed Amazon toward building a delivery network it could control. The company formally launched the

DSP program in 2018 to reduce its reliance on UPS and USPS and to secure more reliable, scalable capacity for last-mile delivery (The Business Journal 2018). Rather than creating a traditional, employee-based delivery company, Amazon shifted much of the operational risk to small contractors, known as DSP owners, while maintaining tight control over speed, performance, and cost structure. This model allowed Amazon to scale rapidly, apply delivery metrics that traditional carriers would never tolerate, and shield itself from much of the liability that comes with managing a massive delivery workforce.

Each DSP operates as an independent business, responsible for managing its workforce and remaining compliant with federal, state, and local regulations. Amazon sets the rules and enforces strict operational standards, but it does not directly employ the drivers, referred to as Delivery Associates (DAs). Instead, DAs are hired and supervised by the DSPs themselves. The DSP handles all employment-related responsibilities, including payroll, scheduling, and benefits. Those benefits, such as medical, dental, and life insurance, are the legal responsibility of the DSP as the employer, not Amazon.

Amazon has intentionally structured the DSP model to retain control over delivery quality while creating legal distance between itself and the people doing the work. By outsourcing last-mile delivery to independently owned businesses, Amazon separates itself from claims related to driver misclassification, workplace injuries, or compliance failures. Any legal or regulatory issues involving drivers fall on the DSPs, not Amazon. That distance does not mean DSPs operate freely. Amazon audits them constantly, often down to the minute, to ensure compliance with the Program Agreement, which governs everything from route assignment to safety enforcement. These audits allow Amazon to maintain high delivery standards while avoiding the risks associated with directly managing drivers. The result is a carefully managed balance. DSPs remain independent on paper, while Amazon retains firm control over how the work is actually performed.

To keep last-mile delivery reliable, Amazon pairs the DSP program with Amazon Flex, a gig-based workforce that fills in the gaps when demand spikes. This hybrid setup gives Amazon access to a flexible labor pool while still relying on the consistency that DSPs bring to the operation. To help DSPs get up and running, Amazon also offers an exclusive vehicle leasing program through a third-party provider. The goal is to make fleet acquisition easier and less financially overwhelming for small business owners. Instead of traditional leases that require hefty down payments, Amazon's program uses a refundable security deposit, and the monthly rates are structured to be competitive. It's designed to let DSPs build and maintain a fleet without taking on the kind of upfront financial burden that usually comes with starting a delivery business.

The available fleet includes a mix of Amazon-branded vehicles designed for different types of delivery work. Smaller vans are used for residential and urban routes, while step vans and custom delivery vans are built to handle heavier package volume. Amazon has also rolled out electric vans as part of its sustainability efforts, giving DSPs a cleaner option that still includes the same telematics and route-optimization features. Every vehicle in the leasing program comes outfitted with telematics systems that track driver safety in real time, capturing data on speed, braking, and overall driving behavior. These tools let Amazon monitor performance across the network without directly managing the drivers, reinforcing its focus on safety while still maintaining the legal separation between Amazon and the DSP workforce.

During peak seasons, especially the holiday rush, this leasing program becomes even more valuable. Its flexible terms let DSPs scale their fleets quickly to handle the surge in demand without taking on long-term financial commitments. Amazon also helps keep those vehicles on the road by offering preventive maintenance services that cut down on downtime, along with a fleet management portal that gives DSPs analytics on vehicle performance, fuel use, and routing efficiency. By simplifying fleet logistics and reducing day-to-day operational headaches, Amazon frees DSPs to focus on the essentials: delivering packages efficiently, keeping safety standards high, and giving customers a smooth, reliable experience.

The DSP model fits neatly into Amazon's larger strategy of optimizing logistics while minimizing direct exposure to labor-related risks. By giving DSPs sophisticated tools, structured oversight, and financial incentives, Amazon keeps its last-mile network among the most efficient in the industry. At the same time, the arrangement raises real questions about where responsibility truly lies. DSPs carry the legal weight of being employers, yet Amazon still controls almost every aspect of how deliveries are carried out, blurring the line between independent partnership and corporate oversight. Whether this setup represents genuine small business ownership or simply another arm of Amazon's tightly managed infrastructure depends on how you look at it. What's undeniable is that the system is built for speed, efficiency, and an almost obsessive drive to optimize every step of the delivery process.

And I was there from the beginning, a moment that quietly set the course for my entire trajectory at Amazon in the years that followed.

DSP Operational Challenges and Solutions

Driver safety is a cornerstone of the DSP program. From hiring through daily operations, Amazon equips DSPs with tools to recruit, train, and monitor drivers effectively. Prospective drivers complete a risk assessment and an extensive background check during the hiring process, with evaluations designed to surface potential safety concerns early. Once hired, Delivery Associates receive onboarding that combines interactive online modules with in-person training sessions. To drive continuous improvement, Amazon tracks driver behaviors such as braking, speeding, and cornering through real-time monitoring systems. Weekly performance scores are shared with drivers, and short, targeted training modules are assigned to address specific issues. These measures are intended to promote safety, but they also reinforce a culture of accountability and constant feedback. I remember one driver who struggled with excessive speeding early on. After a few weeks of focused coaching and corrective training, their performance improved dramatically, a clear example of how effective the system can be when applied consistently.

The operational challenges DSPs face are not abstract or theoretical. They surface daily in the gap between Amazon's expectations and the practical realities of running a delivery business at scale. Route volumes fluctuate without warning. Staffing decisions are made under constant pressure. Equipment, vehicles, and people are pushed to perform at the edge of their limits while standards continue to tighten. The solutions that follow were rarely handed down from above or developed with full visibility into day-to-day conditions. Instead, they were built on the ground by DSP owners and managers responding to problems that had already begun to erode morale, disrupt performance, or compromise safety. In many cases, those solutions were not about optimization, but survival.

Vehicle management is another critical component of the operation. Drivers are required to complete Daily Vehicle Condition Reports (DVCR) or Daily Vehicle Inspection Reports (DVIR) at the beginning and end of each shift to flag maintenance or safety issues. DSPs are also responsible for ensuring their vehicles meet Amazon's branding and appearance standards. When they do not, whether due to skipped inspections or failure to maintain a professional appearance, Amazon can ground the vehicle until the issues are corrected. I have seen many situations where an entire DSP route was forfeited for failing to meet these requirements. Losing a route carries immediate financial consequences for the DSP, and I was often the leader responsible for making that call. To offset some of the operational pressure, Amazon negotiates exclusive agreements for DSPs on fuel programs, insurance, and other essential services, giving them limited tools to manage costs within an otherwise rigid system.

Uniform compliance is mandatory for delivery associates, who are required to wear Amazon-branded attire. This policy ensures professionalism and safety while addressing early challenges faced during the program's rollout. I remember my experience in Portland, Oregon, where the absence of uniforms in the program's early days led to significant misunderstandings. In rural areas, unbranded drivers were often mistaken for trespassers, with one associate reporting that they heard gunfire while making a delivery. Implementing the uniform policy resolved these issues, instilling trust in customers and reinforcing Amazon's professional image.

Operating as a DSP comes with its own set of challenges. These businesses run seven days a week, fifty-two weeks a year, with almost no downtime. DSP owners are responsible for hiring and training drivers, maintaining their fleets, overseeing daily route execution, and staying compliant with Department of Transportation regulations on driving hours and rest periods. Amazon's Central Operations (CO) team also helps DSPs by monitoring routes in real time and intervening when problems arise. Amazon's Central Operations, or CO, plays a major role in keeping everything on track. They monitor driver progress in real time and coordinate rescue routes, along with Amazon's personnel, whenever schedules start to fall behind.

I remember one situation where heavy traffic caused a driver to miss several key delivery windows. CO stepped in immediately, rerouting nearby drivers and working with Amazon's on-site team to offload packages and redistribute the workload. By the end of the shift, every package still made it to its destination. And then there was the peak season in Denver, when heavy snowfall brought deliveries to a crawl. CO didn't hesitate. They redistributed packages across surrounding routes, assigned rescue drivers, and kept the entire operation moving despite the storm. That level of support is essential. It's what keeps routes alive, maintains operational continuity, and ensures customers get their orders even when everything seems stacked against you.

DSPs also face a variety of external challenges that can significantly disrupt their operations. One such issue is the frequent delay in receiving updated vehicle registration tags from state DMVs. When tags expire, the affected vans must be grounded, which can lead to dropped routes, further revenue losses, and negative impacts on performance metrics. Compounding these difficulties is Amazon's tendency to change program requirements without notice. These sudden shifts often force DSPs to scramble to adapt, creating operational chaos and additional strain on their already stretched resources.

During peak seasons, DSPs routinely struggle to maintain staffing levels capable of absorbing surges in demand. High attrition among Delivery Associates is a constant pressure point, driven in large part by the physical strain of the work, with many drivers expected to deliver well over 300 packages in a single day. Turnover is not an anomaly. It is a predictable outcome of a job designed around sustained intensity. In response, Amazon offers recruitment support through standardized job posting templates and access to its proprietary hiring platform, but these measures rarely address the underlying conditions driving departures.

Over time, it became clear that DSP success depended far less on formal support than on adaptability. The system favored those who could absorb instability, improvise under constant change, and keep operating even as the margin for error narrowed. Survival was not about having the best resources. It was about enduring the pressure long enough to stay standing.

DSP Resilience

Daily delivery operations bring a steady stream of recurring problems that rarely show up in planning models. Issues like "Business Closed" or "Address Not Found" routinely disrupt routes, forcing drivers and DSPs into follow-up work that eats into already compressed schedules. Missorted packages create another layer of friction, placing items on the wrong routes and leaving DSPs with an unwelcome decision. Either return the packages to the station and absorb the hit to performance metrics, or ask drivers to deliver them after completing their assigned routes.

Many DAs were understandably resistant to taking on extra work at the end of a long day, but I often encouraged DSPs to deliver missorted packages post-route. The approach was unpopular, yet it produced measurable results. It increased per-package earnings, protected Delivery Completion Rate scores, and helped prevent avoidable customer escalations. It also illustrated a recurring pattern in the DSP program. Problems were rarely eliminated. They were managed by asking people to stretch a little further.

Resilience in the DSP program is often framed as strength or entrepreneurial grit. In reality, it is endurance. DSP owners and their teams are not rewarded for stability. They are rewarded for their ability to absorb disruption, adapt to shifting expectations, and keep operating in conditions that rarely slow down long enough for recovery.

The DSP program's financial structure is built with flexibility at its heart, offering Delivery Service Providers the ability to adapt to the ever-changing demands of the delivery industry. With a range of payment models, including compensation for scheduled block hours, per-piece delivery rates, and options to adjust or relinquish monthly lease fees, the system is designed to provide financial agility. This setup becomes especially critical during peak seasons like the holidays, when DSPs need to scale their operations rapidly to meet surging demand, while still maintaining stability during slower periods.

By tailoring its financial incentives to the unique challenges of last-mile logistics, Amazon supports DSPs in balancing operational efficiency with profitability. This adaptability ensures that DSPs can align their revenue strategies with demand fluctuations, enabling them to manage their workforce and fleet effectively. Additionally, by coupling these payment structures with advanced operational tools and support systems, Amazon empowers DSPs to maintain high standards of service while achieving their financial goals. This thoughtful balance between flexibility and structure underscores Amazon's commitment to fostering sustainable partnerships within its delivery network.

Amazon's technology for supporting Delivery Associates (DAs) stands out as a key strength, offering tools that streamline operations and enhance accountability. As a Return to Station (RTS) Manager at the one of Amazon's largest stations, I witnessed these systems in action both on the road and during the critical moments when DAs returned to the station. A significant part of my role involved investigating the reasons behind undelivered packages, a responsibility that demanded a careful balance of diplomacy and problem-solving to address issues effectively and maintain operational flow.

When DAs returned with packages still in their vans, my first step was to determine whether the issue stemmed from legitimate obstacles, such as address errors, customer unavailability, or vehicle-related problems, or from less justifiable reasons. In many cases, I encouraged DSPs to send their DAs back out to complete deliveries, especially when they returned with time remaining in their 10-hour shift. This approach improved delivery completion rates while also helping DSPs protect their performance metrics and overall profitability.

However, this wasn't always as straightforward as it seemed. Some DAs consistently found reasons to RTS packages, citing issues ranging from minor address discrepancies to vague claims of safety concerns. These patterns required careful navigation, balancing the need to hold DAs accountable with the understanding that the job was physically demanding and often frustrating. Addressing these situations involved close collaboration with DSPs, using data from Amazon's tracking technology to analyze patterns and identify whether the reasons for RTS behavior were consistent or warranted.

Resilience kept the system moving, but it also concealed how much strain the system depended on to function.

On-Road with DAs

Over time, I made it a personal mission to advocate for the chance to go out on the road with Delivery Associates (DAs) through Ride-Alongs, determined to experience their work firsthand. After persistent efforts, I was finally granted the opportunity, and it wasn't long before Amazon implemented a policy requiring station leadership to spend a day shadowing a DA. While many leaders didn't follow through with this requirement, I took the initiative seriously and found the experience invaluable. It gave me a deeper understanding of the challenges DAs face daily and an immense appreciation for the physically demanding nature of their work.

One of the most enlightening aspects of the Ride-Along was seeing the DA's handheld device, known as a Rabbit, in action. This advanced tool was instrumental in streamlining the delivery process. The Rabbit provided precise routing technology, GPS navigation, and essential tools to document deliveries efficiently. DAs used the device to capture clear, shadow-free photos of delivered packages, a critical step in protecting against theft accusations. It also displayed detailed customer instructions, ranging from simple requests like "Leave the package inside the gate" to more foreboding warnings such as "Beware of Dog."

While dog-related incidents were infrequent, they were no less alarming. During my Ride-Alongs, I heard countless stories of DAs being chased or bitten while attempting to complete deliveries. Whenever a dog-related issue was reported, I took proactive measures to safeguard the DAs. I deprecated the customer's address in the system, rerouting future deliveries to a nearby Amazon Locker where the customer could retrieve their packages. This small but impactful step not only reduced the risk of dangerous encounters but also underscored the importance of prioritizing DA safety in a high-pressure work environment.

Those Ride-Alongs did more for me than just deepen my respect for the DAs. They pulled back the curtain on what their day actually looked like, the small improvisations they made to keep routes moving, and the way Amazon's tools either helped or held them back. Watching them navigate those chaotic morning starts, tight driveways, apartment complexes, and unpredictable traffic gave me a new appreciation for the grind they faced every single day. It also reminded me why leadership needed to stay plugged into the real work instead of relying on sanitized PowerPoints out of Seattle. Seeing everything up close shaped the way I tried to support DAs. It made empathy feel less like a buzzword and more like a practical requirement if you wanted people to do their jobs well.

As the DSP program matured, though, that connection to reality slowly slipped away. Seattle kept rolling out new metrics that looked clean on spreadsheets but made no sense on the ground. Load Out time was the perfect example. What had been a reasonable forty-five minutes dropped to thirty, then eventually twenty, all in the name of speed. It was the kind of decision that clearly came from someone who had never tried loading a van with 300+ packages while racing the clock. The tighter the window got, the more the process fell apart. Drivers simply did not have enough time to load their vans in proper LIFO order, which left them digging through piles of mismatched packages later in the day. Deliveries slowed down, frustrations spiked, and what used to be a structured system slowly turned into a frantic morning scramble. It exposed the growing gap between leadership's high-level decisions and the messy, human work that kept the last mile running.

Through these experiences, I got a clearer view of just how much resilience and adaptability DSPs and their teams had to bring to the table every day. They pushed hard to hit Amazon's tightening expectations while doing whatever they could to keep their own operations running smoothly. You could feel the effort it took to keep morale steady while the goals kept shifting above them. Watching that play out made the tension between corporate targets and real-world logistics impossible to ignore.

It became a reminder that the people on the ground often carried the weight of decisions made far away from the chaos of morning Load Out or the strain of peak season. DSPs had to bridge that gap, adjusting and improvising in ways the spreadsheets never accounted for. That balancing act is still at the core of the DSP program, shaping how teams operate and how leaders try to keep their people supported in a system that rarely slows down.

The Reality Behind Owning Your Own DSP Business

The idea of owning a Delivery Service Partner business is often presented as an opportunity for independence, entrepreneurship, and financial upside. In practice, that promise rarely matches the reality. While DSP owners carry the responsibilities of running a business, they operate within constraints that limit autonomy and concentrate risk. What is marketed as ownership often functions more like managed dependency, where success depends less on innovation and more on the ability to absorb pressure without falling out of compliance.

Money became another source of pressure. It was common to hear about owners draining savings, taking on debt, or stretching personal finances just to keep the operation alive during that first stretch. Staffing shortages, high turnover, and sudden operational adjustments all hit at once, creating a cycle that could feel overwhelming even for seasoned entrepreneurs. Nearly every owner had their own version of the same story. They walked in with excitement and confidence, then immediately ran into a wall of unpredictable labor issues and daily fire drills that tested whatever optimism they had left.

The owners who made it through that first stretch usually talked about it with a kind of battered pride. Surviving those early months meant learning the business on a deeper level than any onboarding packet could ever teach. Yet the difficulty of reaching that point was real. The climb toward stability took a toll on finances, sleep, and sometimes family life. It was a reminder that the opportunity Amazon sold to new DSPs came paired with serious risk, and the people who stepped into the program carried both sides of that reality. That tension between possibility and pressure is part of what defines the life of a DSP owner.

Even beyond the daily chaos, there was another structural pressure baked into the program that rarely got talked about openly. Amazon's requirement that DSPs carry a minimum roster of roughly 30 drivers created a strange tension for these small business owners. On good days, when the station handed out enough routes to keep everyone moving, that headcount felt like a path toward real profitability because the workforce was actually being used the way the model intended. But the moment route volume dipped, that same requirement turned into a heavy financial weight. You still had to maintain payroll, keep drivers engaged, and cover all of the operational costs that didn't magically shrink just because Amazon had fewer packages to hand out.

That kind of fluctuation made stability feel almost out of reach. Owners spent so much time trying to anticipate the next week's demand, especially during slow seasons or whenever Amazon shifted operational priorities without warning. It left many of them stuck in a kind of limbo, carrying the cost of being "ready" for work that didn't always materialize. It wasn't that they didn't understand the logic behind staying staffed for peak demand. It was the unpredictability that made the whole setup feel like balancing on a moving floor, one where every dip in volume came out of their pocket.

For most DSPs, this mismatch between fixed labor requirements and uneven income became one of the program's quietest but harshest realities. It forced owners to absorb financial risks that Amazon didn't feel, even though they were the ones keeping the last-mile machine running. That constant tension between meeting Amazon's expectations and managing their own survival became another piece of the job, another layer of strain that shaped what it really meant to operate inside the DSP program.

A DSP usually didn't see real profit until they were consistently running 30 or more delivery routes. That was the sweet spot where the numbers finally started working in their favor. But the truth was that many weeks didn't look anything like that. Routes could slip to 20, and often below, with no hint it was coming, leaving owners scrambling to keep the entire operation from drifting straight into the red. When things got tight, some DSPs had no choice but to sideline drivers without pay while they waited for volume to bounce back. It wasn't ideal for anyone, but the fixed costs of the business didn't back off just because Amazon pulled back on routes.

When DSPs did hit that 30-route threshold and above, the whole operation clicked differently. Economies of scale finally showed up, and the revenue was enough to cover the nonstop expenses that came with running a fleet, paying a full staff, and keeping everything moving. For many owners, hitting that level wasn't about getting rich. It was about breathing room. It was the difference between barely hanging on and finally seeing a path toward real stability.

The trouble was that route availability never settled into anything steady. It rose and dropped with Amazon's shifting priorities, customer buying habits, and whatever seasonal surge happened to be next on the calendar. Prime Week, Back-to-School, Thanksgiving, Christmas, all of those stretches could drown a station in packages and push some DSPs to run 50 or more routes in a single day. During those peak moments, monthly revenue could climb above one hundred thousand dollars, giving owners a brief look at what the business was supposed to be. But everyone knew those windows were temporary. As soon as the surge faded, the numbers snapped back to normal and reminded them that the highs were the exception, not the rule.

In the early days of the program, Amazon tried to get ahead of demand by leaning on Midterm Planning (MP) and Last Mile Capacity Planning (LMCP). On paper, both tools were meant to forecast customer volume and help stabilize operations. In practice, the projections missed the target more often than anyone liked to admit. Over time, pieces of that responsibility were pushed down to Area Managers who actually knew the day-to-day rhythm of their stations and understood what was really happening on the ground. It did not fix everything, but it made the planning a little more connected to reality.

Once the Peak rush ended, the whole landscape changed. Route counts could collapse overnight, sometimes dropping to ten in a day, and that shift created a painful financial squeeze for DSPs. Owners were still required to hold a full workforce so they could meet Amazon's staffing expectations, even when there was not nearly enough work to justify it. Payroll kept running. Vehicle leases kept draining the bank. Insurance, fuel, and all the other operational costs stayed exactly where they were. The revenue did not. That imbalance forced DSPs to absorb the cost of underused capacity, and for many, it became one of the toughest parts of staying afloat.

These swings exposed just how fragile the DSP model could be, because the success of a small business ended up tied to package volume that the owner could not control. When the volume was high, the operation felt strong and profitable. When it dropped, the same business suddenly looked unstable. That tension, the pull between big potential and unpredictable demand, became one of the defining challenges for owners who were trying to run a disciplined operation while living inside a system they did not control.

Most distribution centers were home to around ten DSPs, though that number changed depending on the station's size and throughput. Some DSPs learned how to scale, position themselves well, and make the most of high-demand periods. Others spent more time trying to stay ahead of costs that never slowed down, even when routes disappeared. The inconsistency itself became part of the job. It forced owners to navigate a business environment that moved constantly, while still hitting Amazon's strict requirements and keeping their teams intact.

Over the years, I watched this duality play out again and again. Some DSPs rode the wave of high volume and turned peak seasons into real profit, finally seeing the upside Amazon promised when they first signed the contract. Others fell into the gaps that opened when route counts dried up, struggling to cover expenses that were locked in no matter how little work they received. That contrast said everything. The model could reward you, but it could just as easily leave you exposed, and the line between the two was rarely something the owner controlled.

Helping DSPs Navigate a System Built Against Them

Despite the pressure built into the DSP program, working with DSPs became one of the most fulfilling parts of my career at Amazon. Over the years, I onboarded more than sixty DSPs, each arriving with a different mix of excitement, hesitation, and unease. Getting them started required far more than logistics or process checklists. It demanded patience and a deliberate effort to build trust inside a system that often felt oversized, opaque, and intimidating to new business owners.

When I sat down with a new DSP owner, I always began the same way: "The only thing that matters to me is that you make money, and I'm going to show you how". Some looked at me sideways because profitability was not what they expected to hear from Amazon. It was rarely a message reinforced by broader leadership, which operated far from the daily realities DSPs faced. I became a bridge between the system's expectations and what owners were actually up against.

Advocating for DSPs meant challenging expectations that favored optics over reality, often at quiet professional risk. To close that gap, I took a hands-on approach and built training sessions that cut through a lot of the myths floating around the program. The goal was simple: give DSPs the real information they needed to survive inside Amazon's system, not the glossy version they heard during onboarding. I spent hours walking owners through the parts of the business that actually mattered, showing them how to read their scorecards, where the trouble spots were, and what levers they could pull to turn things around. Every DSP faced different challenges, so the guidance had to be specific. Sometimes it meant digging into stubborn staffing problems. Other times it meant finding ways to lift DCR scores or untangling whatever operational mess was slowing them down. No matter the issue, I tried to offer solutions that made sense in the real world, something that supported both Amazon's expectations and the long-term health of their business.

Keeping communication open was at the center of everything I tried to do. It got to the point where DSP owners would call my cell on my days off, looking for guidance on whatever problem had popped up that morning. I never minded it. If anything, those calls reminded me how much transparency mattered in a program where owners often felt like they were operating in the dark. By staying available and being honest about what was coming their way, I could give them at least one steady point of contact inside Amazon, someone who understood how isolating the job could feel. The pressure of the program never really let up, but helping owners navigate that maze of expectations gave them something solid to lean on as the ground kept shifting under them. And somewhere in the middle of all that, I realized the work was giving me something too. It gave me a deep respect for the grit and heart these small business owners brought to the last mile, even when the odds seemed stacked against them.

As that trust grew, the relationships did too. What started as operational guidance slowly turned into something deeper. Over the years, many DSP owners let me into their personal lives, something that meant more to me than any metric or performance score. It was a sign of mutual respect, built through late-night problem solving, tough conversations, and the shared relief that came when things finally went right. The DSP program will always be a pressure cooker with more operational hurdles than most people ever see, but it is also built on a kind of partnership that, when it works, can be genuinely powerful. Owners who stick with it and learn how to absorb the shocks can find real success. For me, the chance to stand beside them through those ups and downs, to celebrate the wins and weather the losses, was one of the most meaningful parts of my time at Amazon. It showed me that even in a company obsessed with metrics, real human connection can still find a way through.

First Days Inside the DSP and DA World

My first real experience working closely with DSPs started in Portland, back when the program was starting out and the structure around it was far more relaxed. Load Out, process of loading delivery vans, looked very different in those early days. Delivery Associates had the space and the time to sort their vans the way that actually made sense, using a simple Last In, First Out approach. The totes holding the final stops went up front, and the first stops were stacked in the back, creating a natural flow for the route. It was a system that made an immediate difference on the road. Drivers moved through neighborhoods without circling back, without digging through piles of mismatched packages, and without the stress that came from trying to make sense of a van packed in a hurry. It gave them a rhythm, one that supported their day instead of complicating it, and it showed how effective the work could be when the process matched the reality of what drivers needed.

As the DSP program grew, Seattle tightened Load Out metrics, cutting the window from forty-five minutes to thirty and eventually twenty. The changes favored speed over practicality and showed how far decision-makers had drifted from conditions on the floor and road. DAs no longer had enough time to load vans in LIFO order, leaving packages disorganized and routes harder to complete. Frustration rose, undelivered packages increased, and customer complaints followed. A once orderly process became rushed and chaotic, exposing the growing disconnect between corporate decisions and last-mile reality.

Through these experiences, I saw firsthand the resilience and adaptability required of DSPs and their teams. They worked tirelessly to meet Amazon's ever-tightening expectations while trying to preserve efficiency and morale. It was a stark reminder of how difficult it is to balance corporate objectives with the practical needs of the people doing the work on the ground. That tension continues to shape the DSP program to this day.

Even with all the obstacles the program threw at them, my time working side by side with Delivery Associates ended up being one of the most meaningful parts of my career. Their resilience stood out in a way I never forgot. No matter how many demands Amazon added to their day, they kept showing up with a determination that honestly humbled me. These were the people carrying the weight of Amazon's last mile, the ones who held the entire operation together even when the system worked against them. I took a lot of pride in advocating for them, trying to fix the pieces of the process that made their jobs harder than they needed to be. Whether it meant tackling the mess created by rushed Load Out windows or looking for ways to ease the physical grind of the job, everything I pushed for came from a real respect for the work they put in and the dedication they brought to every route.

Working with the DAs reminded me daily how essential empathy, organization, and teamwork are in a high-pressure environment that never slows down. Watching their commitment to the job, even when conditions were far from ideal, made me acutely aware of the human side of Amazon's massive logistics engine. Those moments shaped how I approached leadership. They taught me that listening and understanding matter just as much as problem solving, and that real support comes from taking action for the people on the front lines. Those lessons stayed with me long after the shifts ended. The relationships we built during those years felt solid enough to last a lifetime, and even now I still get calls from former DAs who want to catch up or just vent for a few minutes. It is a reminder that no matter how metric-heavy the system becomes; the whole operation still depends on the strength and spirit of the people actually doing the work.

The Road Reveals Everything

The lessons that emerged in the field were not abstract takeaways. They were consequences, revealed only after decisions met reality. Apartment complexes posed a constant challenge for Delivery Associates, particularly when elevators were out of service. Heavy deliveries such as cases of water, bulk cat litter, and oversized household goods were routine, and it was difficult to watch drivers haul them up multiple flights of stairs, trip after trip. The physical demands of the job were unrelenting. Many new hires quit after completing a single route, realizing too late that the work was far more grueling than they had expected. Delivery was not a job for everyone. Younger drivers or those with strong fitness backgrounds tended to adapt more quickly, while others found the cumulative physical toll impossible to sustain.

Yet, despite the demanding nature of the job, I had immense respect for the older DAs who took it on and refused to let it break them. They may not have had the raw physicality of their younger counterparts, but they pushed through the exhaustion, proving that perseverance mattered just as much as strength. Some of them even underwent noticeable transformations over time, shedding weight, building endurance, and developing a resilience that few outside of the job would ever fully understand. Seeing these individuals embrace the challenge and grow stronger because of it was nothing short of inspiring.

Not every DA approached the job with the same level of commitment. While some pushed themselves to meet the demands of the role, others found ways to manipulate the system in their favor, avoiding as much work as possible. What many of them failed to realize was that Amazon's technology provided me with real-time visibility into their actions. Every stop, every pause, and every deviation from the route was tracked with precision, making it easy to spot patterns of underperformance. Two cases, in particular, stand out.

Martha was a DA who frequently returned to the station with a large number of undelivered packages, blaming her struggles on closed businesses or claiming that her Shipment Per Route (SPR) was too high to manage. Curious about the discrepancies, I started monitoring her location in real time. It didn't take long to notice a pattern: every Thursday afternoon, Martha's van remained stationary for about two hours at a local nail salon. Suspecting a personal detour for a mani-pedi, I decided to intervene. During one of her salon stops, I called the salon and asked the manager to inform Martha that it was time to get back on the road. The impact was immediate. When she returned to the station later that day, her expression of shock was unforgettable. From that moment on, Martha's delivery performance dramatically improved, and the "Thursday salon detour" was no more.

Then there was Jose, another DA who consistently returned to the station with undelivered packages, claiming his shift was not long enough to finish his route. Using tracking tools, I started monitoring his movements and noticed a pattern. Every day around noon, his van parked at an apartment complex that was not listed anywhere on his route. It did not take long to confirm what was happening. Jose was stopping at his girlfriend's apartment for nearly two hours each day, framing it as his lunch break.

When I confronted him, Jose explained that he was entitled to a 30-minute lunch and two 15 minute breaks, and he promised to shorten his visits. For a couple of weeks, he followed through. Before long, though, he slipped back into the same routine.

One afternoon, during one of his extended lunch breaks, I decided to intervene directly. I called Jose's personal cell phone, and his girlfriend answered in the middle of what was clearly a private moment. Keeping my voice calm, I asked her to pass along a message to Jose. It was time to get back on the road. That interruption did the trick. Jose did not take his lunch, or anything else, at his girlfriend's apartment that day.

From that point forward, his on-the-job performance improved noticeably, and the extended midday breaks stopped altogether.

I continued to utilize this monitoring process over the following years, which proved useful not only for performance management but also for responding to incidents involving accidents or stolen vans. In these cases, we would contact local law enforcement to provide support. While I did receive some complaints from DAs about being overly monitored, one incident stood out: a DA named Sergio reported to HR that I had suggested he use duct tape to wrap a dog bite and continue delivering packages. I don't recall if I actually said that, but it certainly sounds like something I might have done. Interestingly, a few years later, I hired him to play a key role in Amazon's Driver Trainer Program, highlighting the complexities of our interactions.

These experiences showcased the remarkable capabilities of Amazon's technology in managing and optimizing DA performance. Tools like real-time GPS tracking, route visualization, and location monitoring proved invaluable for operational oversight. They allowed managers to pinpoint inefficiencies, address recurring issues, and enforce accountability, ensuring that even the most challenging situations could be resolved effectively. Both cases with Martha and Jose demonstrated how a proactive approach, paired with the strategic use of technology, could transform persistent problems into opportunities for improvement. At the same time, they highlighted the complexities of managing a diverse workforce in a high-pressure environment, where human ingenuity and technological innovation must work in tandem to achieve success.

Technology, no matter how advanced, could not solve every challenge of last-mile delivery. Missorted totes and racks staged in the wrong areas regularly disrupted the Load Out process, turning an already time-sensitive operation into a scramble. Delivery Associates, working against the clock, were often forced to search for misplaced packages, delaying departures and putting tight delivery schedules at risk. These breakdowns led not only to missed on-time deliveries, but also to undelivered packages being returned to the station. Once back, those packages were either reassigned to Flex drivers later in the day or pushed into the next delivery cycle, leaving customer satisfaction and promised delivery dates (PDDs) on unstable footing.

In the early days of the program, this lack of coordination during Load Out was common and created inefficiencies that rippled through every stage of the operation. For many DAs, these frustrations compounded the already grueling demands of the job and pushed some past their limits. High turnover became a persistent problem, forcing DSPs to scramble just to maintain performance metrics and keep pace with Amazon's delivery expectations. These challenges exposed how fragile the system really was. Keeping a network this large and fast moving on track required more than technology alone. It demanded constant attention to process improvement, better coordination on the floor, and a willingness to address the human side of an operation that depended on people as much as systems.

One of the biggest challenges during Load Out was the time Delivery Associates spent retrieving racks and loading them into their vans. Depending on the vehicle type, that often meant moving four to six or more racks, a physically demanding task that regularly delayed departures and strained the station's On Time Dispatch (OTD) metric. The pressure to meet OTD was relentless, and any slip quickly drew frustration from station leadership, which was often directed at the On Road Team and me. To address the issue, I reassigned associates to help pull racks directly to the vans, with priority given to step vans that required the most time and effort to load. The impact was immediate. The workload on DAs eased, Load Out delays dropped, and departures became far more consistent.

The results were remarkable. The new system streamlined operations so effectively that the station began consistently meeting the allowable Load Out time. Before long, the process was running so smoothly that the last van often left the station in under 15 minutes, an achievement that once seemed impossible. Its success was undeniable, and the system was eventually adopted as a standard practice across the network, a testament to its impact on improving efficiency.

While this innovation brought significant operational benefits, the recognition for its success went to another leader, who used it as a springboard for promotion. It was disheartening to see my efforts go unacknowledged, but the experience offered a critical lesson about leadership dynamics and the realities of organizational recognition. It reinforced that meaningful change often comes from those working closest to the challenges and that the value of innovation lies in its results, not in the credit received. Despite the lack of acknowledgment, I took pride in knowing the solution had made a lasting difference, proving that practical, ground-level adjustments could transform even the most demanding environments.

Dispatch Cycles

Dispatch is where planning collides with reality, and where small decisions quietly determine whether a day holds together or begins to unravel. Most Distribution Centers operate around a single morning dispatch, known as the Cycle 1 wave, which typically launches between 5:00 a.m. and 9:00 a.m., depending on region and daily volume. Under normal conditions, that single wave is sufficient to move volume through the building and keep the operation stable. During Peak, however, the structure strains almost immediately. Holidays, Prime Day, and major sales events drive demand far beyond baseline levels, exposing how quickly a system designed for efficiency struggles to absorb sustained surges.

To keep pace, additional dispatch waves are layered into the schedule. A Cycle 0 Wave might start as early as 4:30 in the morning, well before the building usually comes to life, while a Cycle 3 Wave can start closer to late morning around 11:00. These extra waves help spread out the volume so packages keep flowing instead of backing up. Even when demand spikes, the system keeps moving and customers still get their orders within the promised windows.

The duration of delivery routes assigned during these waves can vary significantly, typically ranging from 7 to 10 hours. This fluctuation depends on several factors, including package volumes, delivery density, and the geographic layout of the route. During non-peak periods, routes tend to be shorter and more predictable, as demand stabilizes and follows regular patterns. In contrast, Peak periods bring longer, more demanding routes, with additional packages and stops added to meet heightened expectations for on-time delivery. These extended routes reflect the operational intensity required to manage Amazon's commitment to its customers during high-demand times.

These dispatch cycles were not just scheduling tools. They were commitments the station had to live with once volume, labor, routes, and drivers were locked into motion. Adding waves helped Amazon protect the customer promise during Peak, but it also pushed more complexity into the building. Every extra cycle created another deadline, another handoff, and another place where pressure could land when the plan met reality.

Daily routines at Distribution Centers (DCs) are meticulously structured to maximize efficiency and ensure smooth operations. Before drivers even begin loading their vehicles, standup meetings are held to share safety tips, provide operational updates, and discuss areas for improvement. These brief but focused gatherings play a vital role in setting the tone for the day, aligning teams with operational goals, and reinforcing key priorities. Interestingly, these meetings take place at the DSP's parking lot rather than at the station itself, offering a moment of preparation and team cohesion before drivers' head to the station to begin their work. This setup emphasizes the importance of communication and readiness, ensuring that drivers are informed and equipped to handle the demands of their routes effectively. By fostering a sense of team alignment early in the day, these routines contribute significantly to the overall efficiency and reliability of Amazon's delivery operations.

Dispatch cycles gave the operation the appearance of order, but they also marked the point where flexibility quietly disappeared. Once routes were launched and timelines were set, the system stopped adjusting and started expecting compliance. What followed for the rest of the day was not optimization, but absorption. Decisions made in minutes would be carried for hours by people who had no role in setting them. Understanding dispatch as a commitment, rather than a schedule, mattered before looking more closely at who was expected to live with those commitments once the day moved out of planning and into motion.

Seeing DAs as More Than Just Workers

One of the parts of my time at Amazon that I valued most was the opportunity to build meaningful relationships with Delivery Associates (DA), the backbone of the company's last-mile delivery network. These individuals worked under intense pressure, battling tight deadlines and grueling physical demands, yet their resilience and dedication never ceased to impress me. I made it a personal priority to know their names, learn their stories, and recognize the effort they put in each day. The job was demanding, often thankless, and could easily feel transactional, but I wanted to create moments of connection that reminded them they were more than just numbers on a performance report.

Load Out was always a high-energy time, filled with urgency as DAs prepared to hit the road. The focus was entirely on getting out the door as quickly as possible, but in those moments of organized chaos, I made it a point to walk through, greet drivers with a smile, and offer a few words of encouragement. Those interactions may have been small, but they mattered. In an environment where efficiency was prioritized over everything else, a simple acknowledgment could go a long way in reminding people that their work was seen and appreciated.

For many DAs, this job was more than a paycheck. It was a way to support their families, pay down debt, or build a level of financial stability they had struggled to find elsewhere. I had countless conversations with drivers who opened up about personal challenges, ambitions, and the pride they took in hitting their own goals. Sometimes it was finishing a route faster. Other times it was reducing missed deliveries or simply pushing through a brutal shift without giving up. Those moments offered a raw, unfiltered look at the human side of Amazon's operations. Behind every scanned package, every completed delivery, and every metric the company tracked, there was a person doing their best to improve their own life.

The relationships I built during those early mornings and long nights stayed with me long after I left Amazon. Even now, some former DAs still call to check in, share life updates, or laugh about moments from our time working together. Those conversations continue to remind me that leadership is not defined by numbers or efficiency alone. It is rooted in recognizing the people who carry the work. Trust grows through respect, acknowledgment, and a willingness to see individuals beyond their job titles. Long after the metrics fade, those human connections are what remain.

Those relationships became my quiet counterweight to everything that was broken elsewhere in the system. When policies felt detached and leadership grew increasingly abstract, the DAs grounded me in what the work was actually about. They showed up every day, often exhausted, often underpaid for the toll the job demanded, and still found ways to take pride in what they did. Their commitment wasn't driven by slogans or leadership principles posted on a wall. It came from responsibility to their families, to their teammates, and to themselves. In the end, they were the clearest reminder I had that real leadership lives on the floor, not in dashboards or reviews. The company may have measured success in metrics and delivery promises, but for me, success was knowing I treated people with dignity in a system that too often forgot they were human.

The Dispatch Machine

The Dispatch process, commonly known as Load Out, is the heartbeat of Amazon's delivery network, a finely tuned operation where timing, resource allocation, and adaptability must converge to keep the system running smoothly. For On The Road (OTR) Managers, the day begins with orchestrating a complex interplay of how much volume we could handle, how many drivers were available, what vehicles types were available, how routes were built, and how many packages had to move that day. This role demands not just precision and foresight but also the ability to pivot quickly when challenges inevitably arise. In many ways, the entire delivery network's success depends on the efficiency and resilience of the Dispatch process.

Before dawn, OTR Managers start their day by checking in with the Under The Roof team, the group of managers and associates who sort packages through the night at the Distribution Center. Together they review the overnight volume and pinpoint any issues that could slow things down, especially with the line hauls arriving from the fulfillment centers. Those trucks are the lifeline between buildings, and any delay upstream can roll straight into the morning launch and put pressure on the entire operation.

A delayed line haul can throw the workflow into disarray, often forcing the team to shift into ad hoc sorting just to keep the backlog from getting out of control. In those moments, quick thinking becomes essential. OTR Managers have to work immediately with the Delivery Service Partners to reallocate drivers and adjust routes in real time. If a line haul is behind by hours, they may need to trigger a supplemental dispatch to keep packages moving toward customers without losing efficiency. Every decision becomes a balancing act, weighing urgency against accuracy to avoid the kind of ripple effects that can derail the rest of the day.

Another critical aspect of the OTR Manager's role is ensuring DSPs have properly staffed their DAs for the day. Staffing is non-negotiable; without a fully rostered team, the routing process risks descending into chaos. Managers meticulously cross-check DSP rosters against the projected delivery volume, flagging any staffing gaps that could jeopardize the operation. Unassigned routes can derail the entire schedule, leading to hours of manual rework and delayed deliveries. To avoid this, OTR Managers often contact DSP owners early in the morning, urging them to address staffing shortages before routing begins. These proactive measures are vital, as the ability to solve staffing issues quickly can mean the difference between a smooth operation and a day filled with inefficiencies.

At the center of the operation is the routing process, the real brain behind dispatch. This is where the system assigns specific routes to each Delivery Associate and pieces together the path the day will follow. The software maps out the delivery geography almost like a snake moving through a maze, stitching stops together in a way that creates the most efficient route possible. It is a structured process, but it also runs on a tight clock, and even a minor delay can send problems cascading through the rest of the day.

OTR Managers stay focused during this stage, watching closely for anything that looks out of place or could cause trouble once drivers hit the road. Their role is more than fixing issues as they appear. It is about protecting the fragile rhythm of the operation so that every package has a clear, predictable path from the station to the customer's door.

One recurring challenge in the routing process is the misassignment of routes to the wrong DSP or outside the designated Regional Geographic Unit (RGU). These errors demand immediate manual intervention to avoid disruptions in the delivery chain. For example, when a route is mistakenly assigned to a DSP in another region, OTR Managers must rely on tools to reallocate the route to the correct DSP. This process requires swift coordination with the affected DSPs and adjustments to vehicle assignments, often under intense time pressure. These corrections not only test the precision and problem-solving skills of OTR Managers but also underscore the critical role they play in maintaining operational flow.

The ultimate objective of routing is straightforward: to ensure every package reaches the correct address within the promised timeframe. Despite its complexities and potential for errors, the routing process is a cornerstone of Amazon's delivery operation. OTR Managers, equipped with a combination of advanced technology and quick decision-making, work relentlessly to uphold the high standards of accuracy and timeliness that Amazon customers expect. Their efforts reflect a commitment to excellence that keeps the entire last-mile logistics network running smoothly.

Labor planning sits at the heart of dispatch operations because nothing moves unless the right people are in the right places at the right time. OTR Managers oversee the flow of the entire launch, but they depend heavily on the UTR Associates who handle the hands-on work that keeps the morning running. These are the people moving racks to staging, helping Delivery Associates load their vans, and completing Vehicle Safety Audits to make sure every van is road-ready. Without a well-organized UTR team, bottlenecks would stack up quickly and the entire dispatch rhythm would fall apart.

For OTR Managers, effective labor planning means constantly watching the labor board and making sure enough UTR Associates are scheduled to keep up with the morning volume. When gaps appear, whether from unexpected callouts or a sudden spike in packages, adjustments have to happen fast. Leadership will often offer Voluntary Time Off (VTO) to cut labor and hit staffing metrics, without thinking through how the reduced headcount will ripple through the operation. When that happens, Driver Trainers are often pulled in to cover dispatch tasks, giving the remaining UTR Associates space to focus on their core responsibilities or, when appropriate, still take VTO to help balance labor costs. That kind of flexibility is what keeps the launch from falling apart when the morning takes an unexpected turn.

The ability to shift labor on the fly is what holds the entire operation together. Staffing shortages or swings in package volume can trigger delays that ripple all the way to the customer's doorstep. Through careful coordination and a willingness to adapt, OTR Managers and their teams keep the station moving, even when the day throws challenges at them. The smooth handoff between planning and execution shows just how important foresight and flexibility are in a fast-paced logistics environment. If either is missing, the whole system feels it.

As the morning builds toward Dispatch, OTR Managers depend heavily on Amazon's real-time text communication platform, to keep operations on track. This tool is central to managing the inevitable challenges that arise before Dispatch, allowing for swift and coordinated problem-solving. One common issue involves Delivery Associates (DAs) calling out after routes have already been assigned. In these cases, the OTR Manager must act immediately, manually reassigning the affected routes to other drivers. These adjustments often need to be completed within minutes to prevent delays from cascading through the day's operations.

DSP communication through Amazon's real-time text communication platform is just as important when it comes to managing unexpected van service type changes. These issues come up when a DSP suddenly runs short on a specific van type because of maintenance problems or a last-minute breakdown. When that happens, the OTR Manager has to step in quickly, reshuffling routes to other DSPs or substituting whatever vans are available to keep the launch on track. These decisions require quick judgment and accuracy, because even a small delay can ripple outward and put delivery timelines at risk.

This is where Amazon's real-time DSP communication platform made a difference, allowing OTR Managers to coordinate quickly and intervene before a small disruption became a larger operational problem. The ability to adapt without letting the rest of the operation feel the impact is what separates strong OTR Managers from the rest. Their responsiveness keeps Amazon's delivery network steady, even when the morning throws unexpected problems at them. It is this adaptability, grounded in clear communication and fast problem-solving, that helps maintain the level of reliability customers expect from Amazon's last-mile delivery.

Audits are the backbone of operational quality control, serving as a critical component of an OTR Manager's daily and weekly responsibilities. These audits, each with a specific focus, are essential for maintaining safety, efficiency, and compliance with Amazon's stringent operational standards. On any given day, OTR Managers may perform dozens of these checks, ensuring that every aspect of the delivery process meets the company's expectations.

Vehicle Safety Audits (VSAs) are among the most vital, ensuring that DSP vans meet safety and operational standards before leaving the station. Seasonal audits, like Heat Stress and Cold Stress checks, monitor warehouse conditions during extreme weather to protect workers from harmful temperatures. Department of Transportation (DOT) audits confirm that Delivery Associates operating step vans hold the necessary certifications, reinforcing Amazon's compliance with federal regulations. Uniform audits, while seemingly minor, play a key role in maintaining Amazon's professional image, ensuring that all DAs wear company-branded apparel.

These checks extend beyond the warehouse floor and into the parking lot, where OTR Managers inspect storm drains for blockages and address potential hazards that could disrupt operations. Though each audit targets a specific aspect of the operation, together they create a cohesive system designed to keep everything running smoothly, safely, and in line with

Amazon's high standards. Through consistent vigilance and attention to detail, OTR Managers not only uphold Amazon's reputation but also foster a safer, more efficient environment for both workers and customers alike.

Reporting is a cornerstone of the OTR Manager's role, ensuring transparency, communication, and accountability throughout the operation. Each day begins with compiling and distributing safety on-road violation reports to DSP owners, highlighting incidents that demand immediate attention. These reports create a vital feedback loop, enabling DSPs to address violations promptly and improve driver performance. Simultaneously, wave staging reports are shared with Delivery Associates, providing precise location IDs to help them efficiently locate their racks in the warehouse, reducing delays during Load Out and keeping operations running smoothly.

Real-time updates to the On Time Dispatch (OTD) Tracker remain a constant focus throughout the day. This tool provides a comprehensive snapshot of dispatch progress, tracking key metrics such as DSP assignments, loading durations, and departure times. By closely monitoring these updates, OTR Managers can quickly identify bottlenecks and intervene to prevent disruptions, ensuring operations stay on schedule.

As the shift comes to a close, the focus turns to crafting detailed End of Shift (EOS) reports. These reports document the day's events, celebrate successes, and identify areas for improvement, providing actionable insights for the incoming team. By capturing lessons learned and addressing challenges encountered, EOS reports ensure operational continuity and maintain a seamless handoff between shifts. This commitment to thorough reporting upholds the precision and transparency that Amazon's delivery network demands, fostering continuous improvement and ensuring reliability across the operation.

Beyond the immediate demands of the day, OTR Managers are constantly looking ahead, anticipating the challenges that lie in wait. Forecasting capacity is a vital part of this forward-thinking approach, requiring a keen understanding of package volume trends, DSP headcounts, and vehicle availability. This process involves close collaboration with UTR teams to analyze projected volumes and adjust routes, labor, and resources as needed. During surges, especially during Peak periods, DSP owners are contacted early to secure additional drivers and vehicles, ensuring the station is prepared to handle fluctuations. This proactive planning is key to maintaining the consistency and reliability Amazon customers expect, even in the face of unpredictable demand.

Safety and compliance are woven into every part of the OTR Manager's role. Metrics such as FICO Safe Driving Scores, overdue vehicle maintenance, and Delivered Not Received (DNR) packages are closely monitored to surface potential risks early. When safety issues arise, including grounded vehicles that require urgent repairs, managers are expected to act quickly and decisively. Tools like Driver Vehicle Condition Reports (DVCRs) are reviewed proactively to ensure both drivers and vehicles meet Amazon's strict safety and operational standards. This constant attention to accountability and risk management forms the backbone of smooth and reliable operations.

In the fast-moving world of dispatch, OTR Managers serve as the connective tissue that holds the operation together. Their ability to manage competing priorities while adapting to unexpected challenges often determines whether a day succeeds or unravels. Every shift brings new variables, but through careful planning, clear communication, and a consistent focus on customer outcomes, OTR Managers help deliver on Amazon's promise of speed and reliability. The role demands a balance of foresight and adaptability, one that keeps a complex logistics network functioning under constant pressure.

By the end of the day, it became clear that dispatch was not about control so much as containment. Decisions were made quickly, often with incomplete information, and almost never offered clean solutions. Each choice simply shifted pressure from one place to another, with consequences that surfaced hours later on the road. What looked like coordination from a distance was, in practice, a series of compromises made under constraint. Understanding how those compromises took shape mattered before stepping back to examine the systems that set their limits long before the day ever began.

How Amazon Forecasts Capacity

Forecasting at Amazon was never just forecasting. It was the moment assumptions began turning into obligations. Once projected package volume, DSP capacity, supplemental routes, and station limits entered the planning model, those numbers began shaping what the building would be expected to absorb weeks later. On paper, the process created visibility. On the floor, it created pressure before the packages ever arrived.

Amazon relied on two connected planning processes to manage delivery capacity. Forward capacity planning looked several weeks ahead, using projected volume, station limits, dispatch capacity, available DSP routes, and supplemental delivery options to estimate what each location would need. Near-term capacity adjustment operated closer to the delivery date, allowing planners to respond when actual conditions began moving away from the original forecast.

Forward planning established the longer-range framework. It estimated how much volume a station could process, how many routes delivery companies would need to operate, and how much supplemental capacity might be required. It also helped identify staffing or delivery gaps before they became immediate problems.

The process served a legitimate purpose. If a station historically struggled with seasonal volume, planners could anticipate the increase and begin preparing. DSPs might be asked to recruit drivers, secure additional vehicles, or prepare to operate more routes. Supplemental delivery capacity could be reserved for volume the DSP network was unlikely to absorb.

When planners identified a gap between expected demand and available capacity, action could begin weeks in advance. Recruitment could be accelerated, vehicles secured, and station labor plans adjusted. Ideally, leaders had enough time to prepare rather than discovering on the delivery day that the building could not handle the volume.

The weakness was that every forecast rested on assumptions that could change quickly. Customer demand shifted. Drivers quit. Vehicles broke down. Delivery companies lost staffing capacity. Weather interrupted operations. Trucks arrived late, and upstream facilities sometimes sent volume that no longer matched the plan.

Near-term capacity adjustment was designed to address that instability. It incorporated updated package forecasts, current route availability, labor conditions, delivery performance, and station constraints to determine what could still be changed. Leaders could add supplemental routes, redistribute volume, delay certain packages, or reduce the amount of work released into the delivery network when the station approached its limits.

These adjustments became especially important during disruptions. Severe weather could reduce driver availability across an entire region. A late truck could push thousands of packages into a station after the normal processing window. The sudden loss of a delivery company could leave dozens of routes without coverage. Near-term planning gave the operation some room to respond, but that room was always limited.

At one station, the original forecast showed that fourteen supplemental routes would be needed. Before the delivery day arrived, the station unexpectedly lost a significant portion of its DSP capacity. The revised plan increased the supplemental requirement from fourteen routes to twenty, providing enough additional coverage to reduce the number of packages likely to be left behind.

Without that adjustment, the station would have been forced to search for drivers at the last minute, often at greater cost and with no guarantee that enough people would accept the work. Late offers were especially difficult to fill during severe weather, high-demand periods, or when nearby stations were competing for the same drivers.

Forward planning and near-term adjustment were intended to connect preparation with adaptability. The forward plan established expected volume, staffing, route allocation, and station capacity. The near-term process revised those assumptions as the delivery date approached.

Used well, the two processes could prevent a disruption from becoming a full operational failure. Longer-range projections gave DSPs and station leaders time to prepare, while near-term adjustments helped the operation respond to late trucks, staffing losses, weather events, and unexpected changes in volume.

The systems also played an important role in cost control. Overestimating capacity could create delivery blocks that no one accepted, leaving packages without coverage. Underestimating demand could force the station to seek drivers at the last minute, often at higher compensation and with greater risk of missed delivery windows.

The goal was to align projected demand with labor, vehicles, building capacity, and delivery coverage as closely as possible. When the forecast was accurate, the operation appeared seamless. Packages moved through the building, routes were assigned, drivers arrived, and customer commitments were protected.

When the forecast was wrong, however, the consequences moved downstream. Station teams were expected to process more volume. DSP owners were asked to add routes or drivers with little notice. Drivers absorbed heavier workloads. Managers had to explain why performance fell below a target built on assumptions they had not created.

That was where planning became enforcement. The forward forecast set expectations weeks in advance, and those expectations quickly hardened into requirements. Near-term adjustments provided flexibility, but only within the limits established by the original plan.

Once the projected numbers were approved, they stopped being treated like informed estimates and became commitments the station was expected to meet. If conditions changed, the forecast itself was rarely identified as the failure. The shortfall became an execution problem, and the station, DSPs, managers, and drivers were expected to absorb the difference.

The models did what they were designed to do. They translated expected demand into staffing, route, and capacity requirements. What they could not fully capture was the instability of the real world or the human cost of forcing an operation to meet a plan after its assumptions had changed.

That was the central contradiction of Amazon's capacity planning process. The company used forecasting to manage uncertainty, but once the forecast was approved, uncertainty was no longer accepted as an explanation. The numbers became fixed, while the people responsible for delivering them were expected to remain flexible.

Inside the Return to Station Process

In Amazon's delivery network, the Return to Station (RTS) process is more than the spot where a Delivery Associate brings back undelivered packages. It serves as a critical checkpoint that connects daily execution with the feedback needed for long-term improvement. What seems simple from the outside is actually a structured system designed to maintain accountability, track failure points, and drive operational refinement. It keeps the flow of service intact, documents why deliveries were not completed, and provides insights that help shape future planning and performance expectations. By closing the loop on missed stops and recognizing patterns that need correction, the RTS process strengthens Amazon's ongoing push for efficiency and consistency.

At its most basic level, undelivered packages return to the station when the DA finishes their route. Once they arrive, these packages are either moved into a crash sort, where they are rerouted for immediate delivery through Flex and occasionally through a DSP later in the evening. The goal is to complete these deliveries before 10 p.m., although the reality often stretches much later. Other packages are placed in containers to be resorted during UTR operations and scheduled for the next day's dispatch. Both paths carry additional costs for Amazon and disrupt the customer's Promised Delivery Date, creating pressure on the system to minimize RTS volume as much as possible.

Its importance comes from how many moving parts it brings together. The RTS flow depends on organized physical space, the right staffing, technology that tracks every step, and processes that are followed without shortcuts. When these elements work in sync, the RTS area runs smoothly and helps lift overall delivery success rates. It is a quiet but essential part of the operation, reinforcing quality while helping the network evolve through the lessons gathered at the end of each day.

However, achieving this balance is no small feat. The fluid, ever-changing nature of the delivery environment often disrupts the delicate equilibrium required for RTS success. Despite these challenges, the RTS process remains a critical pillar of Amazon's logistics operations. By combining enhanced feedback mechanisms, real-time corrections, and robust accountability measures, it transforms each return into an opportunity for learning and refinement. As Amazon's delivery network continues to evolve, the RTS process will play an indispensable role in upholding the company's high standards of service and operational excellence.

One of the most pressing challenges in the RTS process lies in the disconnect between delivery execution and effective feedback collection. Delivery Associates (DAs) frequently encounter barriers such as incorrect addresses, access restrictions, or unresponsive customers, all of which can prevent successful deliveries. Historically, the RTS process has struggled to systematically capture these insights, leaving critical information unaddressed. This gap results in repeated delivery attempts to problematic locations, creating a cycle of inefficiency that impacts not only immediate delivery outcomes but also the network's broader operational flow.

The consequences of insufficient feedback collection ripple through the delivery network, leading to wasted resources and decreased morale among DAs. When stations lack accurate data on access issues or outdated addresses, they unwittingly dispatch drivers to locations with unresolved problems, resulting in frustration and avoidable delays. For Delivery Service Providers (DSPs), the absence of a structured feedback mechanism means recurring issues on routes remain unaddressed, perpetuating inefficiencies and escalating costs. The lack of a closed feedback loop hinders both short-term delivery success and long-term operational optimization, ultimately eroding customer satisfaction.

At the heart of the improved Return to Station system is a reworked debrief process that transforms a routine package return into a valuable opportunity for refining operations. Rather than serving as a simple end-of-shift procedure, the debrief allows station personnel to engage in structured but natural conversations with returning Delivery Associates. These discussions go beyond basic check-ins, creating space for drivers to share real challenges they encountered on their routes, from unresponsive customers at gated communities to unclear delivery instructions and unexpected obstacles that slowed them down.

Previously, much of this information was lost in the rush to wrap up the shift, with drivers returning undelivered packages without much discussion about what went wrong. The debrief process changes that by creating an open exchange where delivery barriers are not only documented but also addressed when possible. While the process follows a general structure to maintain consistency, it is designed to feel more like a conversation than a scripted checklist, making it easier for DAs to speak candidly about their experiences.

By shifting the focus from routine package returns to meaningful problem-solving, Amazon is not just collecting data but creating a feedback loop that gives frontline workers a voice in improving operations. Drivers who once felt disconnected from the broader system now have a way to communicate the realities of last-mile delivery, making the process feel less like another procedural task and more like a genuine effort to refine the delivery experience for both employees and customers.

What sets this improved system apart is its focus on real-time problem-solving. During these debriefs, reported issues such as incorrect addresses or access restrictions are immediately entered into Amazon's delivery systems, allowing for on-the-spot updates and corrections. This capability prevents repeated delivery attempts to problematic locations, streamlining future operations and reducing wasted time and resources. Unlike traditional methods, where such updates could take days or even weeks to implement, this proactive approach ensures that issues are addressed before they escalate.

The Return to Station process was not designed to recover a bad day. It was designed to account for it. By the time packages came back, the decisions that shaped those outcomes had already been made, and RTS existed to document, categorize, and assign responsibility rather than to resolve root causes. What looked like cleanup was, in practice, a mechanism for translating operational shortfalls into records, metrics, and accountability. Understanding RTS in that light was essential before examining how its physical layout and constraints further shaped what recovery actually looked like in practice.

RTS Physical Layout and Vehicle Management

I played a big role in shaping how Last Mile handles returns, and one of the contributions I'm still proud of was designing the operational layout for the RTS area. I built it around three things that the process desperately needed at the time: throughput, segmentation, and real process control. The model I put together ended up becoming the standard layout that stations still use today. And just like what I had already experienced earlier in my career, another leader eventually took the process and used it as their Promo Doc to move up the ladder. It was frustrating, but it was also proof that the design worked.

The layout itself was built around two distinct lane types, Pit Stop and Escalation, each one tailored to a different return profile. Pit Stop lanes operated like high-velocity service bays, built to clear standard RTS packages fast, with a goal of getting each vehicle through in under a minute. Those lanes carried the bulk of the volume, absorbing a nonstop stream of packages and totes without slowing the station down. Escalation lanes were designed for everything that didn't fit the clean, straightforward path. They handled the messy stuff: multi-package RTS cases, bad addresses, access issues, and anything that needed extra documentation or system updates. By splitting the workflow this way, the RTS operation finally had a structure that could smooth out the wild swings in volume, prevent queues from backing up, and give the building a predictable rhythm instead of constant chaos.

Vehicle management within the RTS area operates under carefully enforced protocols designed to maintain order, efficiency, accountability, and safety. Delivery Associates are required to stay near their vehicles during the RTS process, ensuring that the flow of operations remains seamless and that vehicles can be moved quickly when necessary. This close proximity not only supports operational efficiency but also enhances security by minimizing the risk of packages being misplaced or left unattended.

As packages came back into the building, we unloaded them into designated carts set up inside clearly marked drive lanes. It was a simple system, but it stopped the usual chaos from taking over. Even during Peak, when the pressure in the RTS area could hit a boiling point, that structure kept the space from collapsing into gridlock. By keeping both the vehicles and the returns moving in a clean, predictable pattern, we dodged the kind of bottlenecks that could ripple through the entire station and ruin the flow for everyone else.

And honestly, that choreography of space and movement was the real backbone of the RTS process. Without it, the station would have drowned under the volume. With it, we could push through huge waves of returns without blowing the timelines for sorting or redistribution. Those protocols weren't just rules on paper; they were the only reason the system functioned at all. When people followed them, RTS ran like it was supposed to. When they didn't, the entire operation felt it immediately. It was one of those parts of last mile where the smallest bit of discipline made the biggest difference.

The physical layout of RTS operations imposed limits that no amount of process discipline could overcome. Space constraints, vehicle flow, and dock capacity shaped what was possible long before performance metrics were applied. When congestion built or recovery efforts stalled, the issue was rarely execution alone. It was the environment itself, forcing tradeoffs between speed, safety, and accuracy. Understanding those constraints mattered before turning to the technology layered on top of them, which promised precision but could not remove the limits of the space it was built to manage.

Turning Friction into Data

The RTS process relies heavily on advanced technology to streamline operations and ensure accuracy in address management and delivery. Central to this system is the Address Management (AM) platform, a powerful centralized interface designed to handle address corrections and delivery instructions with precision. AM allows operators to assign essential attributes to addresses, such as access codes, business hours, property types, and specific delivery instructions, transforming raw logistical data into actionable insights that drive efficiency across the network.

By centralizing and automating critical aspects of address management, the Address Management (AM) system plays a pivotal role in enhancing the efficiency and reliability of Amazon's last-mile delivery operations. This seamless integration of technology ensures that drivers are better equipped with the information they need, routes are optimized for maximum efficiency, and operational bottlenecks are significantly reduced. The result is a smoother and more reliable delivery experience for both customers and drivers, underscoring Amazon's commitment to innovation as a cornerstone of its logistics strategy.

When Delivery Associates ran into problems out on the road, whether it was a Business Closed, Unable to Access, or No Safe Location stop, the RTS team logged those issues into AM right away. Some of the situations were more dramatic than the acronyms made them sound. We had plenty of cases where a DA showed up to a house and got met by a dog that wasn't just unfriendly but flat-out dangerous, and the delivery had to be pulled for their safety. In those moments, AM became more than a database. We used it to deprecate the customer's address and reroute future packages to the nearest Amazon Locker so no one else ended up in the same situation.

That kind of real-time feedback loop was the only way to keep address intelligence fresh and useful. It stopped the system from sending drivers back to the same problem spots day after day, and it helped shape smarter delivery strategies across the board. By capturing what actually happened on the road and folding that information back into the tools, Amazon reduced repeat errors, kept DAs out of unnecessary headaches, and tightened the overall efficiency of last mile delivery.

RTS technology was built to increase visibility and enforce precision, not to introduce discretion. Address management systems, scans, and automated validations reduced ambiguity, but they also narrowed the space for judgment once errors surfaced. When something went wrong, the tools did not ask why. They recorded what happened and locked it into the system. Understanding how technology converted operational friction into permanent data points mattered before examining how those records were later used to assess quality and assign accountability.

RTS Quality Control and DSP Accountability

Performance monitoring within the RTS area goes beyond basic productivity metrics, embracing a broader focus on quality assurance and adherence to service standards. Staff members are trained not only in the technical aspects of package handling but also in the essential soft skills needed for effective interaction with Delivery Associates (DAs) and the resolution of on-the-ground issues. This dual emphasis on technical proficiency and interpersonal skills ensures that the RTS process operates with a high degree of both efficiency and effectiveness, fostering a supportive and solutions-oriented environment.

A significant evolution within the RTS process is the introduction of a robust DSP accountability framework. This system establishes a clear chain of responsibility for delivery performance, emphasizing proactive issue resolution and operational excellence. Through regular performance reviews and systematic tracking of delivery challenges, DSPs receive detailed feedback that highlights operational patterns and areas for improvement. By leveraging this feedback, DSPs are empowered to make data-driven adjustments, ensuring that service issues are not only identified but also effectively addressed.

The accountability framework also includes well-defined procedures for corrective actions when performance deficiencies are detected. Depending on the issue, DSPs may be required to implement additional training programs for their teams, develop comprehensive performance improvement plans, or adjust their operational workflows to align with Amazon's high standards. Success metrics are closely monitored to evaluate the impact of these interventions, creating a cycle of continuous improvement. This structured approach ensures that the RTS process not only maintains operational integrity but also drives ongoing enhancements across the delivery network, reinforcing Amazon's commitment to excellence and reliability.

Maintaining high standards of quality control and compliance is a cornerstone of the RTS process, where every package undergoes a meticulous 100% scan verification. This rigorous approach ensures complete accountability for all items, with any missing packages immediately flagged for investigation. Such precision reflects Amazon's commitment to preventing losses and maintaining trust throughout the delivery network, ensuring that no package is misplaced during the return process.

Documentation requirements within the RTS framework are equally strict. All exceptions, whether tied to undeliverable packages or process deviations, must be clearly documented, with adherence to standard operating procedures audited on a regular basis. This level of record keeping supports both immediate operational needs and long-term objectives by generating data that informs future process improvements and maintains audit readiness. By emphasizing documentation and compliance, the RTS process strengthens day-to-day efficiency while also reinforcing Amazon's broader goals around transparency and continuous improvement.

Quality control within RTS was framed as an objective measure of performance, but in practice it functioned as a mechanism for concentrating accountability. Metrics captured outcomes without context, converting complex operational realities into pass fail judgments that landed most heavily on DSPs. Once recorded, those assessments rarely invited explanation or reconsideration. They became fixed reference points used to justify consequences after the fact. Understanding how quality was defined and enforced in this way mattered before examining the standards that governed daily behavior and left little room for deviation.

RTS Standard of Work

The RTS Standard of Work process stands as a cornerstone of Amazon's logistics operations, providing a structured approach to managing undelivered packages and ensuring their efficient reassignment for redelivery. This process demands detailed coordination, rigorous oversight, and strict adherence to established protocols to maintain the flow of operations and uphold Amazon's commitment to customer satisfaction. Central to this workflow is the RTS Area Manager, whose role is pivotal in orchestrating these efforts throughout the shift.

The Area Manager's Day usually kicks off around noon with a check-in alongside the On Road team to review FLEX capacity and map out the same-day delivery push. Amazon gives customers the option to receive their order the same day if it's placed within a set window, and stations in dense population areas tend to shoulder most of that demand. Both Same Day and Left Behind packages are sorted and assigned to either Flex drivers or DSPs, depending on capacity and timing. That early huddle matters more than it seems because it's where we make sure every undelivered package from the morning Dispatch, the Left Behinds, is accounted for and pulled into the same-day plan. Once we confirm those packages are collected, organized, and ready to be reassigned, the rest of the day has a chance of running the way it should. That bit of proactive cleanup keeps issues from snowballing and keeps the workflow moving, all while supporting Amazon's promise of getting packages to customers on time.

As the day advances, the Area Manager takes part in the 2:30 standup (SUM) meeting with the RTS Associate team, finalizing preparations for same-day operations and ensuring a smooth transition into the RTS shift. By 3:00, the manager leads a standup with associates to outline expectations and align the team on priorities. These structured touchpoints are critical for maintaining clarity and focus, particularly in the dynamic RTS environment.

Throughout the shift, collaboration with Delivery Service Providers (DSPs) becomes a focal point as the Area Manager or Process Assistant (PA) works to prevent packages from returning to the station. Using advanced real-time monitoring tools, they track drivers' delivery progress and develop Rescue Plans for delayed routes. This strategy involves rerouting drivers nearing route completion to collect and deliver packages from behind-schedule drivers, ensuring timely redelivery and minimizing inefficiencies. By integrating technology and proactive coordination, the RTS Standard of Work process exemplifies Amazon's dedication to operational excellence and adaptability in its last-mile delivery network.

The manager's role in the afternoon is all about keeping the moving pieces of the station aligned. They oversee the same-day sort, monitor RTS activity, and manage the flow of operations to make sure everything stays on track. One of the more hands-on tasks is scrubbing packages, where undelivered items are reviewed, prepped for sorting, and any oddities or discrepancies are cleaned up before they become bigger problems. The manager also has to make sure updated package data is sent to Central Operations (CO) by 4:00 pm so CO can initiate the routing process for a FLEX dispatch at five or later.

If the volume spikes, it's not uncommon to schedule multiple FLEX sorts later in the evening to keep the backlog from creeping into the next day. That level of attention keeps inefficiencies from stacking up, smooths out the workflow, and ensures that every package in the pipeline is accounted for and ready to move with the accuracy Amazon expects.

As the shift moves into the evening, the RTS team settles into a different rhythm, taking on the behind-the-scenes work that prepares the station for the night sort. They test laptops, printers, and sort-lane equipment to make sure everything is operational before the next wave of packages arrives. Once the final sort is complete, they sweep the building for lingering packages and process returned items through a step known as scrubbing, the manual correction of package statuses so the routing system can assign them to the proper destination.

The team then completes the reports required to close the shift, stages routes, resets equipment, positions racks, and logs into laptops so the next operation can begin without delays or surprises. Throughout the process, they remain in close contact with dispatchers to ensure a smooth handoff to the incoming team.

In addition to overseeing FLEX dispatch preparation, the manager takes on a broader range of responsibilities to ensure smooth station operations and compliance. Station and yard audits are conducted regularly to address logistical challenges and uphold operational standards. Throughout the day, the manager closely tracks road performance for all routes while continuing the package scrubbing process for the final RTS dispatch wave.

Depending on the workload and any additional FLEX sorts, the manager initiates Voluntary Time Off (VTO), allowing associates to leave early when operational demands permit. This strategy not only helps optimize labor costs but also ensures that resources are allocated efficiently. Alongside this, the manager diligently corrects any associate time or punch errors, maintaining the accuracy of payroll and attendance records, which is a critical aspect of ensuring smooth personnel management.

As the shift approaches its conclusion around 10 pm, the manager directs the final stages of the RTS process. This involves the meticulous scan-down of all remaining packages, ensuring that every item is accounted for and accurately logged in the system. Once scanned, these packages are staged for the night sort, setting the stage for processing and inclusion in the following day's dispatch. This attention to detail ensures that operations remain seamless, minimizing disruptions and maintaining high service standards.

To wrap up the day, the manager pulls together a detailed handoff that captures everything the next shift needs to know. It includes notes and reports outlining the day's activity, the challenges that came up, and the key wins worth carrying forward. I eventually created a standardized template for this handoff, one that captured every data point tied to the RTS process. That template ended up becoming the model everyone used, and in true Amazon fashion, another leader later repackaged it and used it to boost their own visibility. Still, the structure mattered. A clean, consistent handoff kept the operation steady, letting the next shift step in without losing momentum. By sticking to these practices, the RTS operation maintained the kind of rhythm Amazon depends on, carrying forward that focus on precision and improvement day after day.

The OTR RTS process was positioned as a way to streamline returns, close feedback loops, and correct recurring issues, but in practice it functioned more as a mechanism for control than recovery. While debrief conversations, DSP feedback channels, and address corrections created the appearance of responsiveness, they operated within a framework that prioritized documentation and compliance over context. Improvements were measured by how efficiently issues were captured and categorized, not by whether underlying pressures were meaningfully reduced. Staffing adjustments and workflow refinements helped contain failures, but they did not change the conditions that produced them.

The RTS Standard of Work was meant to ensure consistency, but in practice it left little room for judgment. Procedures were fixed, expectations were uniform, and deviation was treated as failure regardless of circumstance. When conditions changed, the standard did not bend. People had to. That rigidity shaped the lived experience of working inside the operation, where the pace rarely slowed and the system expected compliance.

Rediscovering Purpose in an Unexpected Place

Before Amazon, I had spent years building and leading two technology companies that, at their peak, employed hundreds of people. My work centered on solving problems, developing products, serving customers, and guiding teams through the uncertainty that comes with growing a business.

That experience shaped how I understood leadership. I was accustomed to making decisions, setting direction, and seeing a clear connection between an idea and its outcome. I had also learned that strong operations depended on more than technology. They depended on communication, adaptability, and people believing their work mattered.

Entering Amazon as a warehouse associate placed me at the opposite end of the organizational structure. Instead of leading a company, I was now moving packages on a night shift inside one of the largest logistics systems in the world. That contrast changed how I experienced the work. I did not see the warehouse only as an associate learning a new job. I saw it as a former business owner trying to understand how a system of that scale coordinated people, technology, and pressure.

Now here's the funny part: despite my past experiences as an entrepreneur, I found myself unexpectedly falling in love with this new chapter of my career at Amazon. What began as a stark departure from my previous path evolved into something deeply fulfilling. The role challenged me in ways I hadn't anticipated, blending the physical demands of the job with a constant need to adapt, make quick decisions, and collaborate with people from an incredible diversity of backgrounds. It wasn't just about the tasks themselves, but the entire experience, from the cutting-edge technology that powered operations to the camaraderie and shared purpose among the team, that struck a chord with me.

The physicality of the job was an especially surprising aspect that I grew to appreciate. After years spent in boardrooms or behind a desk, the hands-on nature of this work brought a refreshing change of pace. Each shift left me with a tangible sense of accomplishment, whether I was moving packages, troubleshooting an issue, or helping streamline the flow of operations. Unlike the abstract achievements of my entrepreneurial ventures, the results of my efforts here were immediate and visible. There was something deeply gratifying about knowing that I was playing a key role in ensuring packages reached their destinations on time.

One of the most enriching aspects was the cultural diversity within the team. Working side-by-side with people from all walks of life opened my eyes to different perspectives and approaches. The blend of languages, backgrounds, and experiences created a vibrant, dynamic environment that was both inclusive and inspiring. It was a constant reminder of the global reach of Amazon and the diverse world we live in. This diversity was a gift, not only making the work environment more interesting but also contributed to better problem-solving and creativity as we learned from one another.

Thinking on my feet and making quick decisions became part of the job's rhythm. In the fast-paced world of logistics, there was no time to dwell on challenges. The urgency of each task kept me sharp, requiring me to adapt quickly and execute with precision. The thrill of this fast-paced decision-making brought me back to my entrepreneurial roots, where agility and quick thinking were key to success.

But perhaps the most fascinating part of the job was the technology. Amazon's advanced systems, whether it was the sortation technology or the complex algorithms driving package deliveries, were awe-inspiring. The seamless integration of human effort and cutting-edge tech felt like a glimpse into the future. The way technology enhanced productivity, optimized workflows, and reduced errors was impressive and something I hadn't experienced on this scale before.

This unique combination of factors of the physical challenges, the cultural diversity, the fast decision-making, and the technological innovation made me want more. It sparked a curiosity and excitement within me, as I realized that Amazon operated with a perspective unlike anything I had seen before. Their approach to business was laser-focused on the customer. Every decision, every process, every technological advancement was made with one goal in mind: customer satisfaction.

What surprised me most was how perfectly Amazon's ethos aligned with my own core values. Throughout my entrepreneurial career, I had always believed that the customer comes first. At Amazon, I saw that same belief reflected in everything they did. It was this shared commitment to putting the customer at the heart of every decision that deepened my connection to the work. In many ways, Amazon's business model mirrored the principles I had held dear throughout my career, making this new experience feel both familiar and invigorating.

Rediscovering purpose in that environment was not about inspiration or renewed motivation. It came from understanding the work as it actually was and finding meaning in showing up consistently within its limits. The routines were repetitive, the demands physical, and the recognition minimal, but there was clarity in knowing exactly what was required and who depended on it. That sense of purpose did not change the system, but it changed how I moved within it. Understanding that shift mattered before stepping further into roles where responsibility expanded, expectations intensified, and the distance between effort and outcome became harder to reconcile.

Meeting the Machine

When I first stepped into the Distribution Center in Ashford, my eyes were immediately drawn to the massive, two-story conveyor system affectionately nicknamed "The Big Iron." This intricate machine dominated the floor, a sprawling web of belts and advanced barcode scanners that served as the beating heart of the operation. Its purpose was deceptively simple yet immensely powerful: to transform the chaos of incoming packages from linehaul trucks into a streamlined process, readying each item for the next step in its journey. From the moment I saw it in action, I knew I was standing in the middle of something extraordinary, where human effort and cutting-edge technology intertwined to tackle the monumental task of last-mile logistics.

Watching packages glide across the conveyors, it was easy to feel awed by the scale and precision of the system. Associates like me played a pivotal role, positioned at specific points along "The Big Iron," where we retrieved packages, scanned them with handheld devices, and placed them into carefully labeled tote bags. Each tote corresponded to a precise location on rack shelves, with unique identifiers that allowed for seamless tracking and organization. The handheld devices were more than just tools; they were lifelines, ensuring that every scan and placement was accurate and that no package was misplaced in the labyrinth of the warehouse. The system's design demanded not just technical efficiency but also a high degree of focus and coordination from every associate on the floor.

What struck me most was the rhythm of it all. The conveyors hummed steadily, packages flowing like water through a network of diverters and scanners, while associates moved with synchronized efficiency, scanning, sorting, and staging. It was a mesmerizing blend of human and machine, a choreographed dance where even the smallest misstep could disrupt the flow. Yet, when it worked, and it often did, it was nothing short of remarkable. This wasn't just a facility moving boxes; it was a testament to Amazon's relentless pursuit of precision and reliability, where the promise of speed was not an abstract ideal but a tangible reality made possible by the coordination of countless moving parts.

Over time, "The Big Iron" became more than a machine to me, it symbolized the magnitude of Amazon's logistics network and the sheer ingenuity required to keep it running. Standing amidst the hum of conveyors and the bustle of activity, I came to appreciate the larger narrative at play. This was a system designed not just to move packages but to embody the scale, ambition, and complexity of a company that thrives on solving the impossible. It was here, in the midst of this finely tuned chaos, that I began to see the intricate balance between technology and human effort. A balance that, at its best, felt like witnessing a masterpiece in motion.

Standing there day after day, I slowly realized that "The Big Iron" was not impressive because it was flawless, but because it depended so completely on people to make it work. The machine amplified human effort, but it also magnified human limits. When associates were sharp, rested, and supported, the system moved with breathtaking precision. When they were exhausted or rushed, the cracks appeared instantly. That lesson stayed with me. Long before I understood Amazon's metrics, leadership principles, or network strategy, I learned this truth on the floor in Ashford: no amount of automation can replace the human element. The technology may set the pace, but it is the people who ultimately carry the promise forward, package by package, scan by scan, keeping a massive system alive through sheer focus and effort.

Navigating Cultural Dynamics and Building Alliances

One of the most unexpected and unsettling realities of working in the warehouse was the open presence of multiple gangs among the associates. These groups, while largely sticking to their own circles, wielded an intimidating influence over the environment, often threatening those outside their factions. Their presence created an atmosphere that was both charged and unpredictable, adding a layer of tension to the already demanding work.

I quickly realized that I was not exempt from these dynamics. The constant undercurrent of hostility made it clear that navigating the situation required more than just keeping my head down. Thankfully, my upbringing had instilled in me a strong sense of street smarts, which proved invaluable in reading situations and defusing potential conflicts. I understood the importance of forming alliances, and I sought out individuals I could trust to maintain a sense of security and support in a space that could feel anything but safe. This vigilance and adaptability became key survival tools as I learned to maneuver through the warehouse's social complexities.

After surviving my first week on the night shift, I knew I needed to find allies who could watch my back in the unpredictable warehouse environment. My first instinct was to approach the biggest guy in the building, a towering Black man who exuded an intimidating presence. I made an effort to engage with him, but he wasn't interested. He brushed me off without so much as a glance, leaving me uncertain about my next move. It wasn't until I noticed him walking a mile to a bus stop after every shift that I saw an opportunity to connect.

For a couple of days, I observed him making this trek, clearly as exhausted as I was after the grueling hours. One early morning, I decided to approach him. As he stood at the bus stop, I pulled up in my car, rolled down the window, and asked if he'd like a ride. He gave me a look that could only be described as "WTF" before silently opening the passenger door and getting in. That was the start of an unlikely friendship. His name was Ralph, though I later started calling him Ralphy.

As Ralphy and I spent more time together during our drives, our friendship deepened, and he began to share more about his life. We found unexpected common ground in our childhoods, both shaped by growing up in Black communities, though our paths had taken vastly different turns. Ralphy's life had been particularly rough. His path had been especially hard. He'd spent years in prison, and the reason he rode the bus every day was because he'd lost his license after a terrible accident that landed him in the hospital for months. One day, he pulled up his shirt to show me the scars across his chest and stomach, vivid reminders of his ordeal. It was a sobering moment that reinforced just how much he'd endured.

Despite his challenges, Ralphy had a warmth and intelligence that belied his tough exterior. There was something incredibly humanizing in our interactions, a mutual recognition of each other's struggles and strengths. I think it gave him a sense of comfort that an older white guy didn't just see him as a stereotype but engaged with him as an individual. For me, it was equally rewarding to connect with someone so resilient and empathetic. Those rides were more than just a commute; they were moments of shared reflection and mutual respect.

Ralphy's openness extended beyond our friendship. He introduced me to other Black associates at the station, making it clear that I was "one of them." His endorsement helped me navigate the complex social dynamics of the warehouse and strengthened my relationships with colleagues. That's how I met Shaniqua, a hardworking mother of a four-year-old boy. Her husband worked for a grocery chain, and they coordinated their schedules so one of them was always home to care for their child. Shaniqua's story, like Ralphy's, was a testament to the strength and determination of the people I worked alongside at Amazon. Each person had a story, a set of challenges they were navigating, and a resilience that inspired me.

Ralphy, Shaniqua, and I became an inseparable trio during our time at the station. We made it a point to align our tasks whenever possible, ensuring we were assigned to the same areas and taking lunch breaks together. Those moments became the highlight of the grueling shifts. We'd talk about everything, our kids, health, and whatever was happening in the world. It was a breath of fresh air in an otherwise intense environment. Working alongside them felt less like a job and more like being part of a team that genuinely looked out for one another.

During the Pick and Stage process, we always looked out for each other. Ralphy, with his big personality and even bigger strength, took it upon himself to handle the heavy stuff. He'd wave me off and scoop up the dreaded 50-pound bags of cat litter, the giant sacks of dog food, and those never-ending cases of water like they were nothing. Watching him lift all that while I sidestepped the strain was equal parts relief and proof of how much we depended on each other to get through the shift. And honestly, those items are still the bane of associates and delivery drivers everywhere. I still catch myself thinking, why can't customers go to Costco and haul that stuff themselves? Having Ralphy by my side didn't just make the work easier; it played a big part in me surviving the warehouse trenches at all.

Our bond made the job bearable, but it also made the losses harder. One morning, Shaniqua called me with devastating news, she wouldn't be returning to work due to a heart condition. I remember standing there with tears in my eyes, struggling to process the thought of not seeing her smiling face during those shifts. I made it a point to check in with her occasionally, asking how she was doing and letting her know she was missed. Still, the station wasn't the same without her, and the hole her absence left was palpable.

Looking back, that period taught me something no training module or leadership principle ever could. In a place that could feel dangerous, impersonal, and relentlessly transactional, survival and meaning came from human connection. The alliances I built were not about protection alone. They were about dignity. Ralphy and Shaniqua reminded me that beneath the labels, reputations, and assumptions people carried into the warehouse, there were real lives, real pain, and real strength showing up every night just to make it through another shift. The gangs, the tension, and the exhaustion were all real, but so was the solidarity that formed when people chose to look out for each other. That lesson stayed with me long after I left the floor. In environments stripped of comfort and safety, it is not policy or process that keeps people going. It is trust, loyalty, and the simple act of seeing one another as human beings.

The Chaos of High Volume and the Threat of Violence

Occasionally, I'd be reassigned to a different shift on the same schedule but different days. On those shifts, without my close-knit group of Ralphy and Shaniqua, things felt noticeably lonelier. I chose to keep my head down and focus on my assigned tasks, sticking to the process and avoiding unnecessary interactions. But these shifts could be especially chaotic, particularly during peak times like holidays, back-to-school season, or Prime Day. The sheer volume of packages was overwhelming, and the pressure from leadership to meet time metrics turned the work into an all-out sprint.

During these high-volume shifts, the atmosphere turned chaotic. Leadership, equipped with blowhorns, shouted at associates to move faster, creating a work environment that felt more like a boot camp than a logistics operation. The relentless pressure to stay within strict time metrics often came at the expense of quality. The careful handling of customer packages, a fundamental part of Amazon's promise, was quickly abandoned in favor of sheer speed.

As the urgency intensified, standards unraveled. Associates resorted to tossing packages across the station, and in some cases, smaller parcels in padded envelopes were flung like frisbees to their staging locations. What should have been a streamlined, methodical process turned into a frenzied scramble, where efficiency was measured in seconds shaved off the clock rather than the care put into each order. It was a mess, both in execution and in principle, revealing how Amazon's obsession with speed could sometimes override the very service standards it claimed to uphold.

Inevitably, the chaos would result in someone getting hit by a flying package, leading to tempers flaring and heated exchanges. The floor would often turn into a disorganized disaster zone, with packages strewn everywhere. Picking up and stowing the mess was a royal pain, adding even more stress to an already high-pressure environment. Those shifts, without the camaraderie of my usual team, were some of the hardest to endure. They underscored how much the job depended on both teamwork and a manageable pace. Without either, the work could quickly spiral into chaos.

One particular night, tensions hit a boiling point. A Latino guy was carelessly hurling packages over the conveyor belt into my area, leaving me to pick them up and put them back on the conveyor for resorting. Several packages hit me directly, and I'd had enough. Frustration took over, and I called him out, yelling that he was full of shit and needed to focus on his job so the rest of us wouldn't have to clean up his mess. The second the words left my mouth; I realized what I'd done. He immediately started shouting back, threatening to "Fuck me up."

That's when I noticed he wasn't just any associate, he was the leader of the Latino crew, covered in tattoos and oozing intimidation. He vaulted over the conveyor belt and charged toward me. I was alone, with no backup, and I knew there was no way out. My adrenaline kicked in, and I braced myself for the fight. Every muscle in my body tensed, my temper boiled, and I stood tall, refusing to back down. As he got inches from my face, staring into my eyes, I said, "Fuck you," fully expecting the next moment to involve a fist.

And then, something completely unexpected happened. He stopped, stepped back, and started laughing. "I like you," he said. "You have big Cajones." I was stunned. What the hell just happened? His name was Hugo, and from that point forward, the tension between us evaporated. We ended up working side by side without any further incidents, and I even started to really like the guy. Hugo was a hard worker, and over time, he opened up about his life, telling me stories about his many "Chicas" and the kids he had. We found common ground in humor, joking and laughing together, and making the job a lot more enjoyable.

After that night, I noticed a shift in how the other Latino associates treated me. They started greeting me warmly at the start of every shift and always lent a hand when it came to stowing heavy packages. The change in dynamic was incredible. They welcomed me into their world, inviting me to their potlucks, where I got to enjoy spicy, authentic Mexican cuisine that was absolutely delicious. What began as a tense standoff turned into a genuine camaraderie. I came to see Hugo and his crew as a great group of people, fun to work with, supportive, and full of life. Those moments of laughter and connection became some of my favorite memories from my time at the station.

One night during a lunch break, I found myself sitting alone when Chong, the leader of the Asian gang, unexpectedly sat down beside me. Chong was an imposing figure, towering, heavily tattooed, and radiating a presence that reminded me of Oddjob, the ruthless villain from James Bond movie, Goldfinger. His reputation preceded him, and I'd heard enough stories to know how he dealt with associates who landed on his bad side, usually in the parking lot after shift. As he sat there, I couldn't help but wonder if I had unknowingly crossed some line or offended someone in their group.

To my surprise, Chong broke the silence not with confrontation but with something entirely unexpected. He told me he was impressed by the way I had handled myself in a recent altercation with Hugo, someone who had been causing trouble. He said it showed character and that he respected that. It was a moment of relief, though I stayed cautious. Over time, as we occasionally worked together, I found that respect held steady, and I never encountered any issues with members of the Asian gang. Chong, for his part, remained an enigma, polite but reserved about his personal life, which was fine by me.

With that, the Asian gang was no longer a source of worry, and fortunately, I'd never had any problems with the group some jokingly referred to as the "Crazy Whites" either. By that point, I'd stopped looking over my shoulder on the walk to my car after shifts, finally feeling a sense of ease I hadn't known since I'd started.

Looking back, those nights taught me a hard, unpolished truth about survival in places where structure breaks down and pressure replaces judgment. Respect was not granted by titles, policies, or HR slogans. It was earned face to face, often in moments where backing down felt safer but standing your ground mattered more. What could have turned violent instead became connection, not because the system protected anyone, but because human beings recognized something in each other. In an environment driven by speed, fear, and constant strain, order did not come from leadership shouting through blowhorns. It came from mutual respect, shared hardship, and the unspoken understanding that everyone was just trying to get through the night. Once that respect was established, the chaos softened. I stopped feeling like a target and started feeling like I belonged. And in a warehouse where belonging was never guaranteed, that made all the difference.

Observing Leadership Challenges

As I began to examine Amazon's approach to leadership, I found myself reporting to Lilith and Juno, two shift leaders who had been given enormous responsibility early in their management careers. They were expected to oversee a Distribution Center moving more than a billion dollars' worth of packages annually while leading a large and diverse workforce under constant operational pressure.

Over time, I would work with both of them again in different stations and roles. My first experience with them revealed a broader problem in Amazon's leadership model: the company often placed promising but relatively inexperienced managers into demanding operational roles without enough preparation, authority, or support. The issue was not their potential. It was a system that expected them to perform like seasoned leaders before giving them the experience and resources needed to succeed.

The weight of their responsibilities was overwhelming. Not only were they responsible for managing the intricate logistics of a fast-paced warehouse, but they also had to lead a diverse team of over 100 Associates. These Associates came from various cultural backgrounds and brought an equally diverse range of attitudes and work ethics to the floor. Balancing operational efficiency with the challenge of managing such a broad spectrum of personalities would have been a formidable task for even the most experienced leaders, yet these young women were thrust into the role with little apparent preparation or support.

The job required quick thinking, decisiveness, and the ability to remain calm under pressure, qualities critical for navigating the chaotic environment of a massive warehouse and the relentless demand to meet Amazon's stringent performance metrics. Pete, the station's Senior Leader, was officially their superior, but he was rarely present during my early days, and my interactions with him came much later in my tenure. Left to manage on their own, Lilith and Juno often seemed overwhelmed by the scale and intensity of their roles. When crises inevitably erupted, their decision-making sometimes faltered, as if they were unsure how to handle the mounting pressure or anticipate the ripple effects of their choices. It was a stark example of how Amazon's leadership model sometimes set individuals up for failure by prioritizing rapid advancement over substantive readiness for the job.

That said, my personal experience with them was largely positive. They treated me fairly, and I had no trouble following their instructions. However, many of the Associates held a different view. Complaints about their "bossy" leadership style circulated regularly, with some feeling that they lacked the experience or tact to lead effectively. This highlighted a recurring issue I noticed during my time at Amazon: an apparent preference for placing inexperienced leaders in operational roles, even in positions where seasoned leadership could have made all the difference.

During one particularly tense shift, Juno expressed her frustration to the Associates over their failure to meet the strict 15-second window for Pick and Stow tasks, a complaint that had become routine among leadership. At the time, I did not fully grasp the pressure driving these demands, but I later realized that these rigid performance expectations were tied to the very metrics Juno and her peers were judged by, numbers relentlessly scrutinized by senior leadership. The stakes were high, and the weight of those expectations inevitably trickled down to the Associates on the floor, turning what should have been a structured workflow into a constant race against the clock.

On this occasion, however, one of the Associates had reached his breaking point. Frustrated and visibly irate, he pulled out a $100 bill, slammed it onto the conveyor belt, and challenged, "If you can do it in under 15 seconds, the money is yours." It was less about the cash and more about making a point, a direct callout of the unrealistic expectations being forced on them. Juno's reaction was immediate, she rolled her eyes, shot him a withering glare, and stormed off without another word. There was no attempt to take the challenge, and everyone knew why. No matter how skilled or experienced an Associate was, hitting that kind of speed consistently was nearly impossible, a fact leadership chose to ignore as they pushed for numbers that looked good on paper but had little connection to reality.

Interestingly, this incident seemed to mark a turning point. Not long after, the time expectation for Pick and Stow was adjusted, allowing a more realistic window for Associates to succeed. Whether the change was the result of internal feedback, relentless pushback from the floor, or some broader recalibration, it offered a rare moment where the metrics, for once, aligned more closely with what was actually achievable. It was a small but significant reminder of the disconnect between the lofty demands of leadership and the realities of the warehouse floor.

Looking back, that moment with the hundred-dollar bill captured the essence of what I was beginning to understand about Amazon's leadership culture. The problem was not individual intent. Lilith and Juno were not villains, and neither was the associate who snapped. They were all reacting to pressure cascading downward through a system obsessed with metrics and blind to human limits. Young leaders were promoted quickly, handed enormous responsibility, and judged relentlessly on numbers they had little power to shape, while associates were pushed to perform at speeds that ignored physical reality. When the expectations finally shifted, it wasn't because leadership suddenly gained empathy. It was because the fiction could no longer be sustained. That small adjustment exposed a larger truth: meaningful change at Amazon rarely came from wisdom at the top, but from friction at the bottom, when the people doing the work finally forced reality into the conversation.

Breaking Points

One day, during a shift break, I stepped into the management office area to report a late linehaul arrival. What I encountered was a scene that starkly revealed the toll Amazon's relentless pressure could take on its young leaders. Lilith was curled up in a ball on the floor, sobbing uncontrollably, her voice trembling as she cried out that she couldn't handle the job anymore. Juno stood nearby, trying to console her, muttering about someone being cruelly insensitive while berating Lilith. It was an unguarded moment, raw and deeply human, and I quickly left the office, not wanting to intrude.

At the time, I didn't fully grasp the weight of what I'd witnessed. But with hindsight and through my own experiences, I've come to understand what likely led to Lilith's breakdown. She had probably just endured the Weekly Business Review (WBR), a dreaded conference call involving operations managers from across the region. These calls were infamous for their intensity. Senior leadership often tore into station leaders for failing to meet rigid, predetermined metrics, their criticisms blunt and public, sparing no regard for the personal toll it took on those being chastised.

The metrics driving these calls were rarely grounded in reality, set by individuals far removed from the daily grind of warehouse operations. These numbers ignored the complexities of running a station, overlooking the countless variables that could disrupt even the most efficient plans. When managers failed to meet these unattainable goals, they were met with harsh criticism and public shaming, their performance reduced to nothing more than a single number on a spreadsheet. Leaders like Lilith and Juno were labeled as underperformers, no matter how hard they pushed their teams or how many obstacles they navigated. I have experienced this system firsthand, and it fosters an environment fueled by anxiety, self-doubt, and, all too often, burnout.

This relentless drive for efficiency at all costs takes a profound toll on the people who keep the system running. Lilith's breakdown was a stark reminder of what happens when a company prioritizes metrics over human well-being, creating an environment where pressure outweighs empathy and unrealistic goals overshadow practical realities. Managers like Lilith and Juno were not just expected to hit near-impossible targets but also to lead a diverse team, manage unpredictable logistical challenges, and absorb the emotional strain of their own stress, all while knowing that failure, even when inevitable, would bring scrutiny rather than support.

Looking back on that moment, I cannot help but admire the resilience of leaders like Lilith and Juno. Despite the crushing pressure and the dehumanizing treatment, they endured from upper management, they continued to show up, doing everything they could to support their teams. Their struggles expose the deep flaws in Amazon's corporate structure, but they also highlight the remarkable strength of those who fight to survive within it. It is a testament to their character, but it is also a call for change. A shift is needed toward more compassionate leadership, realistic expectations, and a genuine acknowledgment of the challenges faced by those on the front lines. Without a change in priorities, the human cost will only continue to rise, buried beneath the weight of numbers that fail to capture the reality of what it takes to keep this machine running.

That image of Lilith on the floor never left me, not because it was dramatic, but because it was honest. It stripped away every polished leadership slogan and exposed the cost of a system that confuses pressure with performance and cruelty with rigor. What I saw that day was not weakness. It was the predictable outcome of a culture that demands everything and gives nothing back when the weight becomes unbearable. Lilith and Juno did not fail Amazon. Amazon failed them by placing impossible expectations on inexperienced leaders, then punishing them for collapsing under the load. That moment crystallized something for me. When a company normalizes emotional breakdowns as collateral damage, the problem is no longer individual resilience or capability. The problem is the system itself, and until that system values people as much as it values metrics, scenes like that will keep repeating behind closed doors, quietly, until someone finally breaks.

Operational Blind Spots No One Talks About

As my journey at Amazon progressed, I found myself gaining a deeper appreciation for the intricate sortation processes that powered the company's logistical backbone. This growth was largely due to Amazon's practice of encouraging Associates to rotate through various roles within the sortation operation. For those willing to embrace the change, this approach offered a valuable opportunity to develop a comprehensive understanding of how the operation functioned as a whole. However, not everyone shared this enthusiasm. Many of my colleagues clung to the familiarity of a single, repetitive task, preferring the comfort of muscle memory and the mental autopilot it afforded.

The value of cross-training became increasingly clear to me as I rotated through these roles. Those of us who chose to engage with the broader operation developed a holistic perspective that improved not only our individual performance but also the overall efficiency of the station. By understanding the interconnected nature of each role, we were better equipped to anticipate and address potential problems before they could cascade into larger issues. It was a mindset that stood in stark contrast to the preference for routine, which often came with unintended consequences.

Over time, I noticed a troubling pattern among some coworkers who sought refuge in the monotony of their tasks. This comfort zone, while seemingly harmless, sometimes enabled behaviors that directly jeopardized the operation's integrity. Some Associates, lulled by the predictability of their duties, began to engage in substance use during their shifts. The signs were unmistakable. Glazed eyes, erratic movements, and a noticeable drop in consistency all surfaced, most clearly during the critical Pick and Stage operations. These behaviors posed serious safety risks and, more critically, undermined the accuracy and reliability of the sortation process.

The consequences of these lapses showed up quickly in one of the operation's most unforgiving metrics: the Promised Delivery Date, or PDD. This was not just another number on a dashboard. PDD represented Amazon's explicit promise to the customer, the expectation that a package would arrive exactly when it was supposed to. That promise sat at the core of customer trust and, by extension, Amazon's dominance. Hitting it required precise coordination across dozens of interdependent steps, from sorting and staging to load out and final delivery. When even a single package was missorted, that coordination unraveled. What seemed like a small error on the floor could cascade through the system, delaying routes, stressing drivers, and ultimately compromising not just one customer's experience but the facility's overall performance.

These moments made the tension impossible to ignore. Amazon chased flawless execution through aggressive goals and tightly engineered systems, yet those systems were only as reliable as the people running them. Designed for precision, the operation still hinged on human judgment, attention, and endurance, leaving it exposed to the smallest slip. For those of us who leaned into cross-training, that reality sharpened our sense of responsibility. The work stopped being just a role to fill and became a point of accountability, a quiet awareness that a single decision or missed detail could steady the entire operation or send it off balance.

Every role, no matter how seemingly minor, contributed to the larger goal of customer satisfaction and brand reliability. A single missort in our facility wasn't just a small error; it could cascade into a late delivery, a disappointed customer, and a potential hit to Amazon's reputation for dependability. This delicate balance between operational precision and human performance was a lesson I came to appreciate deeply. The smallest breakdown in this chain carried the weight of larger implications, underscoring the critical importance of attention to detail in every task.

Looking back, that period taught me that Amazon's greatest strength and greatest vulnerability were one and the same. The system was engineered for speed and precision, but it depended on people being fully present, accountable, and supported to function as intended. Cross-training wasn't just about flexibility or coverage. It was about ownership. When associates understood how their actions affected everything downstream, the work carried meaning and consequence. When they didn't, the cracks spread quickly. That tension stayed with me because it revealed a fundamental truth about operations at scale. No amount of technology can compensate for disengagement, and no metric can replace human judgment. The operation lived or died in the space between process and people, and learning to see that clearly changed how I understood responsibility, performance, and what it really takes to keep a promise moving through a system this vast.

Leadership and Recognition

One of the most challenging dynamics I witnessed during my time at Amazon was workforce management, specifically the constant struggle to balance the system's demand for mechanical efficiency with the need to treat employees as individuals. The operation was engineered for precision. Performance metrics governed nearly every decision, shaping how work was assigned, measured, and corrected. Beneath that relentless pursuit of efficiency, however, an unresolved question lingered in the background. How do you preserve humanity inside a process designed to function with machine-like exactness? How do you keep people motivated and mentally present when their work is repetitive by design, offering little room for creativity or variation? These questions were never fully answered, but they hovered over daily operations, creating a tension leadership confronted continuously, and not always successfully.

That tension became most visible when aggressive performance targets collided with morale and the basic need for belonging. Associates were expected to hit demanding metrics day after day, yet the monotony of the work, combined with constant pressure, often reduced their experience to feeling like interchangeable parts in a massive system. Leadership was tasked with bridging that gap, keeping people engaged while still enforcing the unforgiving standards of speed and precision the operation required. In practice, this balance was fragile. There were moments when leaders genuinely tried to connect, to motivate, to remind employees that they mattered. There were also moments when the numbers took precedence, and the individuals behind them faded into the background. Those moments exposed a deeper vulnerability in the culture, one where results were rewarded far more consistently than well-being, and where the cost of that imbalance quietly accumulated over time.

This experience taught me lasting lessons about leadership, initiative, and the real purpose of process improvement. I saw firsthand how essential it is for leaders to empower associates while also genuinely recognizing their contributions. Innovation and operational progress can look impressive on paper, but they quickly lose their meaning when the people doing the work feel invisible or undervalued. A culture that demands high performance without reinforcing fairness, respect, and inclusion risks alienating the very workforce it depends on. Those observations reshaped how I understood leadership, especially in environments where pressure is constant and expectations never let up.

As I advanced in my career, these lessons became central to how I approached management and problem solving, particularly within the unforgiving pace of logistics and e commerce. They were not abstract principles but practical realities, learned through navigating complexity, absorbing setbacks, and watching how teams respond under sustained pressure. I came to understand that leadership is not simply about hitting targets or optimizing systems. It is about creating conditions where people feel supported, trusted, and capable of sustaining performance over time, even when demands are relentless and margins for error are thin.

My journey at Amazon revealed the delicate interplay between people, technology, and leadership. It highlighted the undeniable strengths of a system engineered for scale and speed, while also exposing the fractures that form when efficiency consistently outweighs humanity. It was a journey marked by challenge, growth, and moments of hard-earned clarity. Those experiences reshaped my perspective on operations and leadership, reinforcing a belief that has stayed with me since. No matter how advanced a system becomes, its success will always rest on the people who power it.

As I rotated through different operational processes, I gradually began to stand out to station leadership. About four months into my tenure, I was offered a role as an Ambassador, a training position responsible for onboarding new Associates and reinforcing safety protocols and operational standards. It was a meaningful milestone. The Ambassador role was both competitive and limited, typically reserved for individuals who had demonstrated a strong grasp of day to day operations and the ability to lead by example on the floor. In most cases, each shift had only one Ambassador, which underscored both the trust placed in the role and its visibility within the station.

Stepping into that position significantly expanded my responsibilities. I became deeply involved in training new hires, making sure they understood Amazon's safety expectations and could execute core processes correctly from day one. Beyond onboarding, I worked closely with experienced Associates as well, helping them fine tune their workflows, resolve recurring issues, and navigate the small but constant obstacles that came with the job. The role placed me squarely between frontline Associates and management, giving me a perspective, few others had. It allowed me to see how decisions made in meetings translated to real work on the floor, while also forcing me to develop communication and leadership skills that went well beyond simply following process. It was a formative experience, one that quietly laid the foundation for everything that came next in my career.

The scope of the role was broad and, at times, exhausting. I trained Associates on safety protocols, pallet jack operation, stowing, staging, and Kaizen 5S, the system designed to improve efficiency and maintain organization across the workspace. I also participated in daily leadership review meetings where bottlenecks were dissected, operational failures were discussed, and improvement plans were debated. Like many expanded roles at Amazon, the Ambassador position came with no additional pay despite the added workload and expectations. Still, I leaned into it fully. I was motivated by the chance to share what I had learned, support my coworkers, and prove that I could operate at a higher level. Just as importantly, I saw the role as a necessary step toward becoming a Shift Assistant, and I took pride in the fact that my work was making a visible difference on the team.

During this period, I worked under a variety of managers, many of whom were young and relatively new to their careers. This dynamic created a collaborative and occasionally unconventional work environment. The line between leader and subordinate often blurred, leading to an atmosphere where we frequently learned and grew together. Their youthful energy and fresh perspectives infused vitality into daily operations, while my experience and willingness to take on challenges made for a mutually beneficial relationship. Together, we navigated the complexities of the station, finding solutions and fostering a sense of camaraderie.

Looking back, that stretch of my career marked the point where observation turned into responsibility. Being an Ambassador placed me at the intersection of pressure from above and reality on the floor, where leadership theory either held up or collapsed in real time. I saw how easily systems could grind people down when empathy lagged behind expectations, and how quickly morale could recover when someone took the time to teach, listen, and lead with consistency. The role sharpened my instincts and clarified my values. It showed me that effective leadership is not about authority or titles, especially in high-pressure environments like Amazon, but about trust earned through presence and fairness. That lesson stayed with me as I moved forward, shaping how I approached every new role and reinforcing a belief that progress only matters when the people carrying the work are able to grow with it rather than be consumed by it.

Portland: A New Journey

The drive itself felt symbolic, almost like a physical mirror of the transition happening in my career. As I headed north, I found myself thinking about everything that had led me to this moment. Each mile became its own pocket of reflection, a quiet place to prepare for whatever waited ahead. Crossing into Oregon, I felt a mix of excitement for the possibilities, uneasiness about the unknown, and a steady determination to prove myself in this next chapter. The trip also gave me the time to dive into audiobooks about Amazon's beginnings. I listened to *The Everything Store: Jeff Bezos and the Age of Amazon*, Brad Stone's bestselling 2013 deep dive into the company's rise. It offered a remarkable window into the world I was about to step into, and it made the journey feel even more significant.

Arriving in Portland, I was greeted by the city's distinctive skyline, a blend of modern structures and historic character, and a quiet sense of anticipation for what lay ahead. The distribution center where I would begin this next chapter sat in Northwest Portland, tucked alongside the Willamette River. Its location felt both intentional and understated, close enough to major transportation routes to support constant movement, yet removed enough to feel almost secluded. The river provided a calm backdrop to the steady pace of daily operations, a reminder that even in an environment driven by urgency and precision, moments of stillness existed just beyond the walls. While the facility was smaller than many of Amazon's larger sites, its responsibility for supporting timely and reliable deliveries across Portland and the surrounding areas carried equal weight.

This move represented more than a change in location. It marked a deliberate commitment to growth, both personally and professionally. I had stepped away from the comfort of familiarity, driven by the belief that real progress often requires discomfort and risk. As I approached the station for the first time, I found myself reflecting on the path that had brought me there, the decisions made, the uncertainty accepted, and the aspirations that had slowly converged into that moment.

The road ahead was clearly going to demand more of me. I was settling into a new city, adapting to a different operational environment, and stepping into a role where I would need to prove myself all over again. Yet as I drove the final stretch and the skyline gave way to the riverfront view of the station, I felt an unexpected sense of readiness. This was not simply a career move. It was a conscious leap of faith, a willingness to grow, adapt, and meet the unknown with resolve rather than hesitation.

The move to Portland felt less like a fresh start and more like a relocation of the same pressures into a new setting. The environment changed, the people were different, but the underlying expectations traveled intact. What stood out was how quickly familiarity returned, not through comfort, but through repetition. Understanding that continuity mattered before settling into the role, where the promise of a new journey would soon collide with the realities of the system it operated within.

Portland's Vibrant Culture and Unforgettable Food

Portland turned out to be an experience unto itself, defined by its culture, its people, and most memorably, its food. The city felt like a culinary playground, a dense patchwork of global flavors woven into everyday life. One of its most distinctive features was its stationary food truck pods, clusters of kitchens serving everything from Middle Eastern, Korean, Japanese, and Mexican dishes to European fare and regional specialties from across the United States. The variety was staggering. For around seven dollars, you could get a fresh, organic, shockingly good meal cooked right in front of you, no pretense and no shortcuts.

One food truck located conveniently near the station quickly became a personal favorite. It served the best Al Pastor tacos I had ever eaten, something I never expected to say given my Southern California roots, where authentic Mexican food sets a very high bar. Al Pastor was familiar territory for me. Marinated pork shaved off a rotating spit, layered onto warm tortillas, finished with salsa and citrus. I thought I knew exactly what to expect. This truck proved me wrong. The meat was smoky and tender, the salsas bright and balanced, the tortillas perfectly charred. I found myself going back again and again. And it was not just Mexican food that stood out. Portland had a way of turning even a quick lunch into a small, memorable experience.

One meal in particular still stands out. Near Nordstrom, I stumbled across a food truck selling corned beef sandwiches. For nine dollars, I was handed a towering sandwich that rivaled anything I had seen at Katz's Deli in New York. It was massive, overflowing with tender, flavorful meat, and left me completely full and deeply satisfied. The relaxed, no frills atmosphere of Portland's food truck scene masked just how exceptional the food really was. It was proof that great cuisine does not require white tablecloths or polished dining rooms, only people who care deeply about what they put on the plate.

Beyond the food, the people of Portland were a revelation. Coming from Southern California, where strangers rarely acknowledge one another, I was genuinely surprised by how warm and open everyone seemed. Simple interactions carried an ease I was not used to. People said hello without hesitation, struck up brief conversations, and showed a level of everyday kindness that felt natural rather than forced. Even the drivers stood out. At crosswalks, cars would stop from every direction to let me pass. It was a sharp contrast to Southern California, where drivers might accelerate just to cut you off or splash you for good measure.

The city's quirks were another part of its charm. "Keep Portland Weird" signs were everywhere, and they captured the spirit of the place perfectly. Individuality was not just tolerated, it was embraced. The city felt comfortable being itself, oddities and all, and that authenticity made it easy to love. At the same time, Portland had a sobering side that was impossible to ignore. Its homeless population was unlike anything I had seen before. Tents lined grassy patches near freeways, parks, and street corners, forming small, makeshift communities. Outside my apartment on NW Thurman, a homeless man slept on the concrete. I made a habit of giving him my empty plastic water bottles, a small gesture that felt human in a situation that otherwise felt overwhelming. Whether coincidence or not, my car remained untouched during my time there.

What stayed with me most was the resilience of those living outdoors. Portland winters were unforgiving. I remember how cold it became during loadout, cold enough that a bottle of Gatorade froze solid on my desk inside the station. Enduring that temperature for a few hours was difficult. Imagining surviving it day after day, night after night, was something I struggled to comprehend. Then summer arrived and flipped the script entirely. By June, the city turned hot in a way that felt closer to Palm Springs than the Pacific Northwest. The heat was dry and intense, carried by wind that seemed to drain the air from your lungs.

Portland was a city built on contrast. Unpredictable weather, a vibrant food scene, eccentric culture, and the visible strain of homelessness existed side by side. Yet beneath those extremes, there was a steady undercurrent of warmth that never fully disappeared. From world-class meals served out of food trucks to small, unremarkable acts of kindness, the city carried a quiet resilience. It left an impression not through spectacle, but through complexity.

That contrast carried into daily life. Portland's culture and food offered moments of relief from the pace and pressure of the operation, but the separation was never complete. Enjoyment existed alongside long shifts and persistent expectations, creating brief pauses rather than escape. Those moments mattered not because they offset the work, but because they revealed how narrow the space for balance had become. Recognizing that tension was important before the focus returned fully to the demands of the role and the operational realities that continued to shape each day.

From Associate to Leader

My promotion to Shift Assistant, later retitled Process Assistant, marked a meaningful turning point in my Amazon journey and my formal entry into leadership. It was the moment my role shifted from simply executing the work to helping guide it. I was now positioned at the center of daily operations, responsible not just for my own output but for ensuring the shift held together the way it needed to. The role expanded my scope and accountability, pushing me further into leadership and reinforcing that I was someone the team depended on to keep things running smoothly.

The responsibilities of a Shift Assistant were wide ranging and relentless. I was responsible for allocating labor effectively, leading start up and mid shift meetings, assigning job duties, and providing clear direction as conditions changed throughout the day. One of the most demanding, and ultimately rewarding, parts of the role was troubleshooting problems in real time. Whether it was a workflow bottleneck, a staffing imbalance, or an unexpected disruption on the floor, I had to assess the situation quickly and adjust without losing momentum. Communication became critical. I needed to stay connected with Associates, peers, and leadership alike, making sure expectations were clear, priorities were aligned, and the operation stayed on track even when things did not go as planned.

Beyond these broader duties, my role often required hands-on involvement in specific operational tasks. I frequently assisted with production duties, trained Associates, and ensured strict adherence to Amazon's processes. One of the most gratifying parts of the job was removing barriers for Associates, empowering them to perform more efficiently while reducing defects. This wasn't just about addressing immediate problems; it was about anticipating issues and establishing systems that prevented them from occurring in the first place.

As a Shift Assistant, I was also responsible for closely tracking performance metrics, ensuring that my area consistently met both productivity and quality targets. Assessing Associate performance was a significant part of my role, requiring detailed and constructive feedback to help team members grow and excel. Using independent judgment, I directed the team toward our objectives while addressing inefficiencies, safety concerns, and other challenges head-on. My intimate knowledge of operational processes made me a reliable point of contact for resolving day-to-day issues.

This role also positioned me as a critical link between Associates and leadership. By providing performance feedback to leadership, I contributed to fostering a culture of accountability and continuous improvement. Maintaining a full understanding of production goals and regularly reviewing process documentation were vital aspects of the job, ensuring that workflows remained streamlined and effective.

Safety quickly became a central pillar of my responsibilities. It extended far beyond hitting productivity targets and into the responsibility of ensuring a safe, properly equipped environment for everyone on the floor. I became actively involved in safety initiatives, identifying potential hazards and offering real time feedback to Associates to help reinforce safe behaviors before issues escalated. When job related injuries occurred, I took immediate action to document and report incidents in accordance with company policy. These moments reinforced how seriously safety was treated and how much accountability rested on frontline leadership to protect the people doing the work.

My leadership responsibilities continued to expand as trust in my capabilities grew. I was asked to provide vacation coverage for Area Managers, stepping into higher level oversight when needed. That responsibility was a clear signal that leadership viewed me as someone capable of maintaining operational stability, making sound decisions, and supporting teams beyond my formal role. It required a broader perspective and a deeper understanding of how decisions at the top of the shift impacted everything downstream.

The Shift Assistant role was demanding, fast paced, and deeply formative. It gave me the opportunity to sharpen my leadership skills, gain a more complete understanding of Amazon's operational engine, and contribute in ways that directly affected both performance and safety. It also reinforced just how critical the role was within the broader system. Shift Assistants were the connective tissue of the operation, holding together people, process, and execution to keep the machine moving efficiently, even under constant pressure.

The move from associate to leader did not come with clarity or certainty. It came with added responsibility layered on top of the same constraints, pressures, and expectations that already existed. The work shifted from execution to decision-making, but the margin for error narrowed rather than expanded. Leadership was less about authority than accountability, often without the tools or flexibility to match it. Understanding that shift mattered before stepping fully into a role where success would be measured not by effort, but by outcomes shaped by forces largely outside of one's control.

Perks on Paper

One of the role's biggest advantages was Amazon's benefits package, which reflected a genuine investment in employee wellbeing. The pay increase was modest, a pattern I would see again with later promotions, but the benefits helped offset it. The medical coverage was especially strong, with multiple comprehensive options that gave me confidence my family and I were well protected.

That sense of security extended to dental coverage, which was equally generous and thoughtfully structured. Knowing that these essential needs were thoroughly covered provided a level of stability that mattered. It allowed me to stay focused on the work in front of me and on my own professional growth, without the constant background concern of navigating healthcare uncertainties.

Beyond health-related benefits, the 401(k)-retirement plan offered a competitive way to plan for the future, helping to build a sense of financial stability over the long term. Perhaps most unique were the stock options, which created a sense of shared ownership in the company. Knowing I had a stake in Amazon's success fostered a deeper connection to the work I was doing and the organization's broader mission. These offerings went beyond the basics, underscoring Amazon's commitment to not just supporting its employees in the present but also investing in their long-term well-being. It was a benefits package designed to inspire loyalty and security, reminding employees that they were valued contributors to the company's continued growth and success.

Another standout feature of Amazon's benefits package was the time off policy, which offered 160 hours of vacation time along with an additional 80 hours of paid time off. On paper, it suggested a meaningful commitment to rest and balance, reinforcing the idea of a culture that understood the value of stepping away to recharge. The structure implied that motivation and productivity were best sustained when employees had space to invest in their lives outside of work, a message that aligned neatly with the company's public emphasis on wellbeing.

The reality, however, often told a different story. The pace of the work, the constant pressure to hit metrics, and the unpredictability of daily operations made taking time off far more difficult than the policy suggested. Coverage gaps, staffing shortages, and the unspoken expectation to remain available all worked against the idea of true disconnection. What looked like a stabilizing benefit in theory often felt inaccessible in practice. The contrast between what was promised and what was realistically achievable became another quiet tension of the job, one that revealed how easily well-intentioned policies could be undermined by the demands of the environment they were meant to support.

Looking back, that period in Portland carried a quieter lesson that didn't fully register until much later. While I was taking on more responsibility, solving real operational problems, and keeping the floor from collapsing under Peak pressure, I was also being paid noticeably less than others doing comparable work. At the time, I didn't question it. I was focused on the job, on proving myself, on keeping things moving. With distance, it's hard not to suspect age played a role. On paper, Amazon's perks painted a picture of balance, support, and long-term investment in its people. In reality, most of those benefits existed somewhere beyond reach. Compressed schedules made them difficult to use, time off bent to operational need, and flexibility rarely showed up when it was actually needed. The problem wasn't what was offered. It was how little of it survived contact with real pressure. Seeing that gap early mattered, especially as I moved closer to leadership, where those same perks were often used to justify expectations that left no room to meaningfully benefit from them at all.

A Unique Station and Seasonal Challenges

Working at the Amazon distribution center serving the Portland area was a distinct and constantly evolving experience, shaped by the station's physical design, operational cadence, and the demands of the Pacific Northwest's seasons. The facility itself was compact and highly efficient, reflecting Amazon's emphasis on streamlined execution. Four large bay loading doors formed the lifeblood of the station, in near constant motion as delivery vans from multiple Delivery Service Providers cycled through. With only a single entrance and exit for DSP traffic, vehicle movement was tightly controlled. That structure allowed leadership to closely monitor arrivals, load times, and departures, reinforcing the precise rhythm required to keep the operation moving at speed.

At the center of the building sat the sort conveyor, the operational core around which everything revolved. Positioned squarely in the middle of the station, it functioned as the primary sorting point where packages were prepared for final delivery. Rows of bins filled with delivery bags extended outward from the conveyor, each one carefully staged for pickup. These bags contained everything from small, lightweight parcels to heavy, oversized items, and the constant motion around the belt gave the space an energetic, almost frantic feel. DSP drivers maneuvered their vans in a steady loop around the conveyor, parking briefly to load, then pulling out and heading straight to their routes. The flow was intense but deliberate. Watching it unfold felt like observing organized chaos, a tightly choreographed system where every movement mattered and small delays could ripple outward quickly.

Portland's seasonal extremes added another layer of complexity to the job. Winters were unforgiving, with cold air creeping into the station and turning routine tasks into endurance tests. Loading, unloading, and sorting required heavy layers, and staying in motion became essential just to keep warm. Summers brought a different kind of strain. Heat radiated off the pavement and metal vans, transforming each shift into a physical grind. These seasonal shifts shaped how we worked and how we paced ourselves throughout the year, influencing everything from staffing considerations to personal energy management. Despite the conditions, a sense of camaraderie developed among the team. Adapting together to the weather became part of the shared experience, and the operation continued to run regardless of what the season threw at us.

In my new role, I worked what was known as the back half schedule, four ten hour shifts from Wednesday through Saturday. My day started early, with shifts beginning at 3:00 a.m. and ending around 1:30 p.m. The hours demanded discipline and resilience, especially during periods of extreme cold or heat. Adjusting to such an early start was difficult at first, but over time it became routine. There was even a quiet satisfaction in finishing my workday while most of the city was just getting started, creating a rhythm that felt distinctly my own.

One advantage that made the schedule manageable was my proximity to the station. I lived just a mile away in a small studio apartment, which meant my commute was short and predictable. The apartment itself was modest, but it offered a quiet space to decompress after the constant movement and noise of the workday. Being that close eliminated the stress of long drives and gave me more time to rest and recover. That balance, limited as it was, played an important role in helping me sustain the physical and mental demands of the job.

Looking back, that station in Portland became more than just another assignment. It was a place where environment, design, and human endurance collided every single day. The tight physical layout, the unforgiving weather, the early hours, and the constant pressure to keep vans moving left little room for error or complacency. Yet within that constraint, I learned how adaptability, routine, and proximity to the work could make even the most demanding conditions survivable. The station taught me that operations are never abstract. They are lived experiences shaped by geography, architecture, climate, and the people forced to make all of it work in real time. That understanding deepened my respect for the frontline realities of logistics and reinforced a truth I would carry forward. No matter how efficient a system looks on paper, it is always the human element that determines whether it holds together or quietly starts to fray.

Lessons from Remarkable Leaders

During my time at Amazon, I had the opportunity to work under several exceptional leaders whose influence left a lasting impression on my professional journey. Each brought unique perspectives and leadership styles, shaping the station's culture and driving us toward success. While not every leader I encountered met this standard, the standout individuals truly embodied the core values that Amazon espoused in its earlier years: empowerment, collaboration, and an unwavering dedication to customer satisfaction. Their guidance not only enriched my experience but also played a pivotal role in my growth, reinforcing the importance of strong, empathetic leadership in such a demanding environment.

Kyle, the Station Operations Manager, was a natural leader with a robust personality. His voice carried through the entire station, and his presence was always felt. He had an uncanny ability to manage the hectic environment with ease, bringing a calm and organized approach even during the most chaotic shifts. What I appreciated most about Kyle was his willingness to invest time in explaining Amazon's intricate operational processes. He didn't just oversee operations; he ensured his team understood the why and how behind every process. One memorable instance was when he took me through the complicated routing process, often referred to as the "snake of routing." His thorough explanations made complex processes clearer, allowing me to become more efficient in my role. Kyle struck that uncommon balance between authority and approachability, and it made him one of the few best leaders I encountered during my time at Amazon.

Aaron, an Area Manager and Army Captain, was one of the most exceptional leaders I had the privilege of working with during my time at Amazon. His background immediately set him apart. During his military service, he flew Blackhawk helicopters and carried with him stories from missions in Afghanistan that were both humbling and awe inspiring. After returning home, he continued serving through the National

Guard, flying air search and rescue missions around Mount St. Helens. Those experiences shaped the way he led. His discipline, calm under pressure, and sense of responsibility were evident in everything he did, and they created a standard that others naturally gravitated toward.

Aaron was more than a manager. He was a genuine mentor who invested deeply in the growth of the people around him. He took the time to teach me how to navigate Amazon's complex systems and software, patiently breaking down tools that often felt overwhelming into concepts I could actually understand and use. His guidance was never rushed or transactional. He wanted people to succeed, and he made himself available whenever questions came up. Beyond the technical side, he shared leadership lessons drawn from real experience. Sometimes that came in the form of direct advice. Other times it came through stories from his military service that reinforced the importance of teamwork, adaptability, and staying composed when stakes are high.

What truly set Aaron apart was his humanity. He balanced professionalism with genuine care in a way that felt rare. He held people to high standards, but he never lost sight of the fact that they were people first. His mentorship had a lasting impact on me, not just professionally, but personally as well. He once shared that his long term goal was to fly for a national airline, a dream he pursued with the same quiet determination he brought to leadership. I was proud to later learn that he achieved that goal and now flies Airbus A320s for United Airlines. Aaron remains one of the finest individuals I have ever met, and his influence continues to shape how I think about leadership, resilience, and what it truly means to lead well.

Looking back, leaders like Kyle and Aaron represented the best of what Amazon could be when leadership was grounded in competence, clarity, and genuine care for people. They didn't lead through fear or performative intensity. They led by understanding the operation, explaining it clearly, and showing up consistently for their teams. Working under them reinforced a lesson that stayed with me long after those roles ended. Strong leadership isn't about volume, rank, or pressure. It's about creating stability in chaos, teaching instead of posturing, and helping people grow rather than burning them out. In an environment as demanding as Amazon, leaders like that didn't just make the job better. They made it sustainable, and they set a standard that was, unfortunately, far rarer than it should have been.

Unlikely Mentors in a Sea of Chaos

Bobby was an interesting character, to put it mildly, and he worked alongside me as a peer. When I first started, he was assigned to provide additional training on the software tools and walk me through the day to day operation. From the outset, our interactions were strained. Bobby's personality was unconventional at best. He was blunt, unapologetically direct, and at times outright abrasive. Early on, I found him to be a bit of a dick, which only made the already steep learning curve more frustrating than it needed to be. His dismissive tone and lack of patience tested me, and finding any kind of common ground felt like work in itself.

Over time, though, something shifted. As we continued working together and were forced into repeated collaboration, the friction began to ease. We learned how to operate around each other, then with each other, and eventually the working relationship settled into something productive. To my surprise, that eventually turned into a genuine friendship. Bobby never completely lost his sharp edges, but I started to understand them. Beneath the rough exterior was someone deeply knowledgeable who took real pride in knowing the systems inside and out. Once trust formed, he was generous with what he knew, and I came to respect both his competence and his strange, unpolished way of showing it.

Despite our rocky start, Bobby ended up having a big impact on my growth. He was the one who helped me get a handle on Amazon's software tools and the data behind them, often offering insight that went far beyond the mechanics of the systems. He pushed me to think on my feet, adapt fast, and approach problems with a no-nonsense mindset. Looking back, a lot of my ability to survive and navigate such a fast-paced, demanding environment came from what I learned working with him. Our shift from friction to friendship was unexpected, but it was one of the more meaningful turns in my early Amazon journey.

Alex, affectionately nicknamed "The Bear," was the night sort Area Manager. With his easy-going nature and great sense of humor, he quickly became one of the more popular figures in the station. Alex brought a lighter tone to the often-intense work environment. One of the more memorable moments with him involved his playful side. One early morning, as I arrived at the station, I was greeted by an 8-foot teddy bear constructed entirely out of Amazon boxes. As I approached, the sculpture suddenly moved, and to my surprise, Alex had hidden inside the bear to greet everyone with a laugh. His ability to make work fun while ensuring things ran smoothly was one of his standout traits.

Alex stood out not just for his sharp sense of humor, but for his ability to teach in a way that actually stuck. One of the most valuable skills he passed on was how to manage the routing process, a critical function that determined whether the station started the day smoothly or immediately fell behind. Rather than simply explaining it, he pushed me to take ownership of it. More often than not, he would insist that I run routing myself as soon as I walked in at 3:00 a.m., forcing me to engage with the process before I had time to hesitate.

That approach was uncomfortable at first, but it was exactly what I needed. Being thrown into the responsibility sharpened my instincts, built confidence, and gave me a deeper understanding of how tightly coordinated the station's logistics really were. Alex did not just teach by instruction. He taught by trust, and that made all the difference.

One ethos Alex instilled in me was the importance of setting the next shift up for success. He firmly believed that every shift should finish its tasks completely, ensuring that the following team wasn't left scrambling to catch up. It was a practice rooted in Amazon's early culture of collaboration and efficiency, one that fostered mutual respect and accountability across shifts. Sadly, as the company evolved under Andy Jassy's leadership after he took over the CEO role from Jeff Bezos, this principle of setting the next shift up for success became less of a priority. However, Alex's example left a lasting impression on me, and I continued to carry that mindset forward in my work.

Alex also made sure our mornings started with efficiency. He ensured that all the Pick Lists for my associates were printed and ready before the shift began. Pick Lists were essential tools, detailing which totes were assigned to specific routes and where to stage them in the warehouse. By having these lists prepared, we could hit the ground running each morning, avoiding delays and maintaining momentum. Associates on my shift would each take on several Pick Lists, organizing and staging the totes for Dispatch. Alex's thorough preparation set a high standard, reinforcing the importance of planning and teamwork in achieving operational excellence. His influence extended beyond technical skills; it was a lesson in leadership, responsibility, and leaving things better than you found them.

On one occasion, Bobby voiced his dissatisfaction with the way the Pick Lists were sorted, making it clear he wasn't happy. The following morning, Alex, never one to let complaints go unchecked, handed Bobby a ream of blank paper and informed him that he would now be responsible for printing the lists himself. The decision added an extra hour to our already packed schedules, and as expected, Bobby struggled to get it right, creating more frustration than solutions.

Despite the chaos that ensued, Alex's approach reflected his unwavering commitment to teamwork and accountability. While his methods could be unconventional, they carried a deeper lesson: if you're going to criticize, you should be ready to contribute to a better solution. This experience left a lasting impression on me, highlighting the importance of ownership and collaboration in a demanding environment. It was yet another example of how Alex's leadership style balanced humor, discipline, and a clear focus on fostering responsibility among the team.

Looking back, Bobby and Alex represented two very different but equally formative lessons in leadership and growth. Bobby taught me that competence does not always come wrapped in warmth, and that friction, when navigated honestly, can sharpen skill and resilience in ways comfort never will. Alex showed me the opposite side of that coin. He proved that trust, humor, and preparation could elevate an entire operation without stripping it of its humanity. Together, they shaped how I learned to operate under pressure, how I handled responsibility before I felt ready for it, and how I began to understand leadership not as a title, but as a daily set of choices. Long before promotions or expanded roles entered the picture, those lessons quietly took root, setting the tone for how I would show up long after those early mornings and night shifts were behind me.

Talent Recognized, Potential Realized

Diana was another exceptional person I had the privilege of working with. Bright, energetic, and quick to learn, she stood out almost immediately as someone with real potential. In her initial role as an Ambassador, she was responsible for coordinating the staging process during loadout, a position that required attention to detail and constant movement. She handled it with confidence and consistency, but it was clear early on that she was not content to stay within the narrow boundaries of that role.

What set Diana apart was her curiosity. She wanted to understand how the entire station functioned, not just her piece of it. Before long, she began asking questions about the technology tools I used every day. These were not simple systems. They were critical to managing bulk deliveries, correcting package statuses, and assigning routes accurately under tight time constraints. Rather than being intimidated, she was eager to learn how they worked and how decisions made inside those tools rippled across the operation. That hunger to grow, combined with her ability to absorb information quickly, made it obvious that she was capable of far more than her current role suggested.

During our breaks, I took the opportunity to mentor her, walking her through the intricate functionalities of these tools. Diana's enthusiasm and quick grasp of the complexities of routing and delivery assignment were impressive. She was a natural, picking up the nuances of the tools and quickly applying them to her work, which made her an invaluable asset to the team. Her ability to adapt and her willingness to go above and beyond her role left a lasting impression on me.

After I left the Portland station, leadership reached out to ask about Diana's performance. Without hesitation, I gave her a glowing recommendation, confident in her abilities and potential. Not long after, she applied for a part-time Shift Assistant role, transitioning seamlessly into the position and excelling. Eventually, Diana moved into a full-time Shift Assistant role, filling my old position. Watching her rise within the company was a testament to her drive, dedication, and talent. Diana's success story remains one of the most rewarding experiences of my time at Amazon, and it's a reminder of how impactful mentorship and shared knowledge can be in fostering growth and opportunity.

Diana's progression reinforced something I had already begun to believe deeply by that point. Talent reveals itself quickly when curiosity is encouraged and knowledge is shared without gatekeeping. She did not advance because someone handed her an opportunity. She advanced because she asked better questions, did the work to understand the system, and showed up ready to carry more weight. In a company that often talked about growth while quietly restricting it, watching Diana move forward on merit alone felt rare and meaningful. Her success wasn't just hers. It was proof that when people are trusted, taught, and taken seriously, they will almost always rise to the level they're capable of, and often far beyond what their job title ever suggested.

Grills, Bonds, and Jäger Bombs

One of the more surprising and genuinely enjoyable parts of the job was the Saturday night BBQs, an unspoken but deeply ingrained tradition among the station's leadership team. Every Saturday night, we gathered at one manager's home to mark the end of a long, punishing week. These were not casual drop ins or halfhearted get togethers. They were a release valve. A place to decompress, laugh, and reconnect as people rather than titles. The food was always on point, usually grilled chicken, fish, and burgers, and the atmosphere felt loose and unguarded in a way the station never allowed.

What made those nights stand out was how inclusive they were. It was not unusual for senior leaders from Seattle to make the three hour drive down just to be there. Titles faded quickly once the drinks came out. Conversations flowed freely, and the usual hierarchy softened into something closer to genuine connection. Alcohol was very much part of the ritual. One of the unofficial rules was that everyone had to do at least five Jäger Bombs, which almost guaranteed that by the end of the night, most of us were well past buzzed and into questionable decision making territory.

Thankfully, Diana did not drink and willingly became my designated driver. Her reliability in those moments probably saved me more than once, especially in an environment where excess was treated as part of the bonding experience. Those nights were not just about blowing off steam. They strengthened relationships with peers and leadership in ways that mattered once we were back inside the pressure cooker of the operation.

That culture existed squarely during Jeff Bezos' tenure. It was widely understood that gatherings like these were encouraged, and in many cases quietly supported, with expenses often covered. There was an underlying recognition that morale and camaraderie mattered in high stress environments. As leadership changed and priorities shifted, that culture slowly disappeared. The BBQs were phased out, another casualty of a growing emphasis on efficiency and optics over community. Even so, those Saturday nights remain some of my strongest memories from the early days. They were a reminder that shared experiences can build trust faster than any meeting, and that real bonds often form far outside the walls of the workplace.

Looking back, those Saturday night BBQs captured a version of Amazon that no longer exists, one where leadership still understood that people cannot operate at maximum intensity indefinitely without somewhere to exhale. They were imperfect, messy, and sometimes excessive, but they were human. They created trust that spreadsheets and dashboards never could, and they reminded us that the people running the operation were just as important as the operation itself. When that culture disappeared, something essential went with it. The work kept moving, the metrics kept tightening, but the connective tissue thinned. Those nights didn't just help us survive brutal weeks. They helped us remember who we were underneath the titles, long before efficiency replaced community as the organizing principle.

DSP Beginning Lessons

Portland was where I was first fully immersed in the world of Delivery Service Providers, during the early days of the program when everyone involved was still finding their footing. It was a period defined by rapid learning and constant adjustment as we tried to make sense of a business model that was evolving in real time. The learning curve was steep, not only in understanding new operational processes, but in learning how to work with the wide range of personalities that came with the DSP ecosystem.

The business processes themselves were not the hardest part. On paper, they were logical and methodical, designed to streamline deliveries and plug cleanly into Amazon's broader logistics network. The real challenge was the people. The operators and owners behind each DSP brought their own expectations, management styles, and interpretations of how the business should run. Those perspectives often clashed with Amazon's standards, and just as often with one another. It was my first exposure to what I privately thought of as "schizophrenic personalities," not in a clinical sense, but as a reflection of how wildly inconsistent and contradictory these viewpoints could be. Every owner believed their approach was the right one, and navigating that landscape often felt less like management and more like herding cats within a system that demanded uniformity.

At times, I struggled. Miscommunication and unclear directions were common pitfalls as I worked to balance the needs of the DSPs with Amazon's expectations. But these challenges became invaluable lessons. This period marked the beginning of my ability to work effectively with such dynamic and diverse personalities, a skill that would later become central to my success with the DSP program.

What began as a chaotic and often frustrating experience ultimately became the foundation of my expertise. Portland was not just my introduction to the DSP world. It was the proving ground where the strategies, communication skills, and adaptability I would rely on later were first forged. Every difficult conversation, every misalignment, and every hard lesson contributed to a deeper understanding of how the program truly functioned.

Looking back, those early days now feel like a crucible. The pressure, uncertainty, and constant adjustment forced growth faster than comfort ever could. They shaped how I approached problems, how I worked with people operating under stress, and how I navigated systems that were still finding their identity. Long before any later successes, Portland did the real work of preparing me for what was ahead.

I vividly recall shortly after my arrival to the station, heading to the back during loadout to check on how things were progressing. The area was dimly lit, the soft glow of overhead lights barely cutting through the shadows that cloaked the rows of vans being loaded for the day's deliveries. It was late, and the usual hum of activity felt muted, almost eerie. As I approached, I noticed eight vans with Delivery Associates (DAs) standing nearby, methodically loading their packages. They worked in silence, their movements precise and deliberate, the kind of efficiency that didn't require words.

As I drew closer, their movements stopped all at once. All eight men turned toward me, their eyes locking onto mine with a cold, piercing focus. This was not the casual glance of someone noticing a passing supervisor. It felt deliberate, measured, as if they were studying me, sizing me up in silence. The atmosphere shifted instantly. The air grew heavy with an unspoken tension, and I felt a sharp, almost physical chill move through me. Their expressions were unreadable, but there was something dark beneath the surface, something that made my instincts flare and told me, without question, that I was not welcome.

I tried to break the tension by asking how they were doing and whether they were close to finishing their load. The sound of my own voice startled me. It came out louder than I expected, awkwardly out of place in the suffocating quiet. No one answered. Not a word. Not a nod. Nothing. The silence pressed down hard, heavy and deliberate, like a weight on my chest. I felt the hairs on the back of my neck rise, a primal response I did not have to think about or question. Years of street smarts and hard learned awareness clicked in immediately. This was not a moment to push, posture, or assert authority. This was a moment to recognize the signal and back away.

Their hostility was not overt, but it was unmistakable. I could feel it radiating off them, directed squarely at me, heavy and intentional. I did not linger. After what felt like far longer than it actually was, probably only a few seconds, I turned and walked away, heading back toward the load out desks. As I put distance between myself and that group, I felt an invisible weight lift from my shoulders, even though a low level of unease stayed with me.

That encounter stuck with me for days. I later learned that the men all worked for the same DSP and were from Afghanistan. Their reputation at the station was already well established. They were not interested in small talk with management, and they made no effort to appear approachable. Whatever tension existed there ran deeper than normal workplace friction. It was shaped by experiences and perspectives that went far beyond the walls of the station or the scope of our day to day interactions.

From that point on, I made a deliberate choice to give them space. It was not driven by fear, but by an understanding that boundaries had been clearly drawn. In environments like this, you learn quickly when engagement is productive and when stepping back is the wiser move. Sometimes leadership is not about inserting yourself into every situation. Sometimes it is about reading the room, respecting limits, and knowing when leaving well enough alone is the most effective decision you can make. That moment taught me an early and lasting lesson. Not everyone operates by the same rules or motivations, and recognizing that reality can be just as important as knowing when to lead from the front.

That experience in Portland became an early marker of how different the DSP world really was from anything I had experienced before. It reinforced that authority alone means very little in environments shaped by history, identity, and unspoken rules. Trust could not be demanded, and engagement was not always welcome, no matter your role or intent. What mattered was awareness. Reading situations accurately. Knowing when to speak and when silence was the smarter choice. Portland taught me that leadership inside Amazon's last mile was not just operational. It was deeply human, cultural, and situational. Success depended less on enforcing structure and more on understanding where structure stopped working. That lesson stayed with me long after I left the station, shaping how I approached people, power, and conflict in systems far more complex than any process map could capture.

Portland: Lessons from the Field

One of the most memorable and wild experiences of my time in Portland was the day I volunteered to ride along with a Delivery Associate (DA) to better understand what they faced on the job. Leadership had put out a call for volunteers, and I eagerly stepped up, ready for an adventure. Little did I know just how unforgettable the day would turn out to be. I was paired with Amka, an Alaska Native Eskimo whose joy and enthusiasm were infectious from the moment we met. Amka wasn't like most people; her energy transformed even the most mundane tasks into something engaging and fun.

From the start, her expertise was obvious. As we loaded her van, she walked me through her system for organizing packages with a level of precision that felt almost artistic. Each parcel was placed deliberately, fitting together like pieces of a puzzle until the van was packed efficiently and cleanly. What made the moment unforgettable, though, was not her skill. It was the fact that a full blown blizzard was raging outside. Snow had piled two to three feet deep across Portland, and conditions were bad enough that I questioned leadership about whether it was even safe to send drivers out. The answer I got was blunt and final. It was up to the DA.

I assumed Amka would delay deliveries until the storm passed. It seemed like the only reasonable choice. When I asked her what she planned to do, she smiled without hesitation and said, "Heck no. This is just a little storm. I won't have any trouble driving." Her confidence caught me completely off guard. I was worried about making it out of the parking lot, and she was treating the blizzard like a mild inconvenience. Against my better instincts, I trusted her judgment and climbed into the van, bracing myself for whatever came next.

Once we were on the road, it became clear that her confidence was earned. Amka navigated the icy, snow packed streets with a calm ease, handling the van as if it were a clear summer morning. As we drove, she told me stories about growing up in the remote wilderness of Alaska. She talked about living off the land, hunting for food, and learning survival skills at an age when most kids were still figuring out how to ride a bike. The stories were captivating, equal parts humbling and inspiring. Her resilience, self reliance, and deep connection to nature came through in every detail, and by the time we returned to the station, my admiration for her had only grown.

Curious about her upbringing, I peppered her with questions. What does walrus meat taste like? How do you stay warm when it is forty below zero? She answered every one with patience and humor, never making me feel foolish for asking. With each response, she painted a vivid picture of a life so far removed from my own experience that it felt almost unreal.

For her, extreme cold and endless snow were not obstacles to be endured. They were simply facts of life. She spoke about them casually, even fondly, with a kind of grounded optimism that was contagious. Listening to her, I realized how perspective shapes resilience. What felt overwhelming to me was routine to her, and that quiet confidence settled my nerves in a way nothing else could.

The challenges that day were relentless. At one point, we pulled up to a ranch with a long driveway buried under four feet of snow. I suggested skipping the delivery and coming back when conditions were safer, but Amka wasn't having it. "No way!" she said, grabbing a large package and trekking through the deep snow with an effortless grace that only someone accustomed to the tundra could manage.

Later, we ran into an even bigger test. Ahead of us was a steep hill completely coated in ice and snow. I was convinced that was where the route would end, that turning back was the only sensible option. Amka never hesitated. She backed the van up, pressed the accelerator to the floor, and drove straight at it.

The van slid back down the hill more than once, tires spinning, gravity winning every time. Each time we slid back, my grip tightened on the seat, fully convinced we were about to end up in a ditch. Amka stayed calm. No frustration. No second guessing. She simply reset, adjusted her angle, and tried again. After several attempts, we finally crested the hill. The moment the van leveled out, I let out a breath I did not realize I had been holding.

Watching her push through that moment was genuinely inspiring. Where I saw danger and uncertainty, she saw a problem to be solved. Her determination never wavered, and by the time we drove on, I had a whole new appreciation for what real confidence under pressure looks like.

I also got my hands dirty that day, delivering my fair share of packages. However, being from the South, I was woefully unprepared for the cold. Despite a thick coat, my choice of thin tennis shoes proved to be a rookie mistake. After hours of trudging through the snow, my feet were freezing, but I powered through, determined not to slow us down. By the end of the day, we had delivered every package in the van, returning to the station nearly ten hours later, utterly exhausted but immensely satisfied.

That day wasn't just a ride-along; it was a profound lesson in resilience, grit, and the challenges DAs face daily. Amka showed me that with the right mindset and a fearless attitude, even the toughest obstacles can be overcome. Her strength and positivity left an indelible mark on me, and I'll never forget her or that snowy day. It remains one of the most humbling and inspiring experiences of my career.

Looking back, that ride-along stripped away any remaining abstraction I had about last-mile delivery. Metrics, dashboards, and leadership meetings could never capture what it actually takes to do that job when conditions turn hostile and there is no safety net beyond your own judgment. Watching Amka operate in that blizzard reframed my understanding of capability and courage. She wasn't reckless. She was prepared, grounded, and deeply confident in skills forged long before Amazon ever entered her life. That day stayed with me because it exposed the quiet heroism embedded in the work Amazon often reduced to numbers. Long after the snow melted, the lesson remained clear. The system runs because of people like her, not because of process, pressure, or slogans, but because of resilience earned the hard way and carried without fanfare.

The Bottleneck Breakthrough

As Peak season approached, station volume began climbing fast, signaling the start of one of the most demanding stretches of the year at Amazon. Peak was not a single busy day or even a hectic week. It was a sustained surge in orders that hit multiple times each year, testing every part of the operation. The most intense stretch came during the holiday season, beginning just before Thanksgiving and often extending well into the new year. Additional spikes followed events like Prime Day and the Back to School rush, each triggering sharp increases in package volume that strained capacity almost overnight.

During these periods, stations across the country routinely saw volume jump by more than fifty five percent. That kind of increase left little room for error. Internal teams had to recalibrate staffing, workflows, and throughput almost immediately, while DSPs scrambled to scale their Delivery Associate workforce fast enough to keep pace. Hiring, training, and retention all collided with rising customer expectations, and any weakness in the system was exposed quickly. Peak was where planning met reality, and where the limits of both people and process were tested day after day.

Our station in Portland, already constrained by its limited space and compact design, was hit particularly hard. The layout, with drive lanes circling the central staging area, became a logistical nightmare when volumes spiked. Overflow packages spilled into the drive lanes, creating bottlenecks. DAs, armed with their Route Lists to locate assigned totes, found themselves wading through a chaotic sea of boxes, totes, and other DAs, all competing for space in the narrow aisles. The inefficiency of this process was glaring, leading to extended Load Out times, frayed tempers, and mounting frustration across the board.

Recognizing the need for a solution, I took it upon myself to analyze the process and identify areas for improvement.

After carefully observing the chaos during Load Out, I developed a plan to address the bottlenecks and restore order to the operation. The first step was reorganizing the staging process. By spreading out the staging areas, I reduced the number of DAs assigned to each aisle from four or five to just two. This simple adjustment immediately alleviated congestion, creating a smoother and more manageable flow for both people and packages.

Next, I tackled the overflow of packages clogging the drive lanes by rethinking how totes were stacked and staged. I restructured the staging process to prioritize totes by wave, the scheduled times DSPs entered the station to pick up their routes. Later-wave totes were staged at the far ends of the aisles, while earlier-wave totes were placed on top. This strategic reorganization not only cleared the drive lanes but also made it significantly easier for DAs to locate their totes quickly, reducing the time spent navigating the station.

To ensure consistency, I also adjusted the Pick Lists associates received at the start of their shifts. These lists, which dictate the order in which packages and totes are staged, were updated to align with the new staging process. This integration ensured the operation ran smoothly from start to finish, minimizing confusion and streamlining the workflow.

While the solution wasn't perfect, the results were immediate and undeniable. The changes dramatically improved the flow of Load Out, reducing congestion and allowing the operation to run far more efficiently. The Loading Area Turnover (LAT) metric, which tracks the time DSPs spend loading their vans and exiting the station, saw a significant improvement. The once-chaotic environment became smoother, faster, and more organized. DAs could locate their totes with ease, the station felt less congested, and the overall experience was far more manageable, even under the pressure of Peak season.

Peak season has a way of revealing truths that quieter periods allow to stay hidden. In Portland, the surge didn't just expose capacity limits or process gaps. It exposed how fragile operations become when growth outpaces design. What stayed with me was not the metric improvement or the cleaner drive lanes, but the realization that stability often comes from simple, floor-level thinking rather than top-down directives. Watching the chaos settle into something workable reinforced a lesson that would repeat throughout my career. When pressure is highest, the most effective solutions are rarely complicated. They come from paying attention, respecting how the work is actually done, and making changes that give people room to move, think, and succeed under strain.

Trust Through Action

Despite the success of those improvements, I never received formal recognition for the work. As is often the case in fast moving environments like Amazon, the credit never circled back to where it started. Surprisingly, that did not bother me. Knowing that I had built something functional, something that reduced friction for both my team and the DSPs, was enough. Seeing the system hold up day after day, especially under the pressure of Peak, became its own reward. It was a quiet reminder that the real value of the work was not in acknowledgment, but in the difference, it made for the people relying on it.

As the months passed, I found myself thriving in Amazon's relentless, ever shifting environment, particularly through deeper engagement with the DSPs. What initially felt overwhelming, navigating clashing personalities, unclear expectations, and occasional tension, gradually became one of the most fulfilling parts of my role. My early hesitation around building those relationships faded as I leaned into consistent communication and mutual respect. Over time, confidence replaced uncertainty, and collaboration replaced friction, reinforcing that strong relationships were just as critical to success as any process or metric.

What started as cautious and strictly professional exchanges gradually evolved into something far more meaningful and productive. Instead of reacting with frustration when challenges arose, we approached problems with a sense of teamwork, working through issues calmly and finding solutions that benefited everyone. Just three months into the job, I was both surprised and humbled when DSPs began seeking my advice on improving performance and optimizing delivery efficiency. This level of trust and respect caught me off guard, marking a shift from simply fulfilling my role to becoming someone they viewed as a reliable partner. What had once been routine interactions became genuine collaborations, driven by shared goals and a mutual commitment to making the operation run as smoothly as possible.

The camaraderie I built with the Associates and DSPs an exceptional experience. These relationships were never just about work. They grew into something real, filled with humor, warmth, and mutual appreciation. What started as casual banter on the floor turned into meaningful conversations that fostered a true sense of community. Each day, I looked forward to these interactions, knowing I was part of a team that genuinely cared about one another's success and well-being.

One of the things that stood out most was their generosity. Many of the Associates and DSPs would bring me local Portland cuisine to try, sharing pieces of the city's vibrant food culture with a kind of pride that went beyond just offering a meal. It was not just about the food itself, but about the connection behind it, a reflection of the trust and camaraderie we had built over time. These small gestures made me feel included in a way that extended beyond the workplace, as though I was part of something greater than just the day-to-day grind of logistics.

This sense of connection was never just about completing routes or meeting delivery quotas. It was about people looking out for each other, sharing moments of understanding, and creating an environment where respect and kindness were just as important as efficiency. What could have been a purely transactional working relationship became something deeply personal, shaping my experience at Amazon in ways I never expected. Long after I left, those relationships remained with me, serving as a reminder that no matter how fast-paced or demanding a job may be, the bonds built through trust and respect are what truly matter in the end.

Looking back, that chapter of my time at Amazon solidified something I had been learning quietly along the way. The work that mattered most was never the work that got applause. It was the work that made someone's shift a little easier, a route a little smoother, a long day slightly more bearable. The trust I earned from Associates and DSPs did not come from authority or titles. It came from showing up consistently, listening, and doing the unglamorous work of fixing problems without needing credit for it. In a system obsessed with metrics and visibility, those relationships became my real measure of success. They reminded me that even inside a machine built for speed and scale, human connection still had weight, and that was something no dashboard could ever capture.

Building Connections Beyond the Workplace

On my days off, coworkers would pull me away from the grind and invite me into their lives outside the station. One of the most memorable of those invitations came from an Associate who worked for me, a former newspaper writer from New York who's past I knew nothing about at the time. He invited me to join him and a few friends on a tour through Oregon's wine country. It was only during that drive that I learned about his earlier life in journalism. As we moved through rolling hills and vineyards, he shared stories from his years in New York, covering everything from high profile events to deeply personal human-interest pieces. Hearing those stories reshaped how I saw him. He was no longer just someone I managed on a shift. He was someone who had lived a full, complex life long before Amazon.

What surprised me even more was his depth of knowledge about wine. This was not a casual hobby. He understood the craft, the history, and the intention behind each bottle, and his enthusiasm was infectious. Until that day, I had no idea Oregon had such a strong winery scene. Suddenly, I was moving from vineyard to vineyard, tasting wines that genuinely impressed me. Bold Cabernet Sauvignons, elegant Pinot Noirs, smooth Merlots, rich Shirazs, and punchy Zinfandels all came with stories he eagerly shared. Each stop felt less like a tasting and more like a lesson, delivered with excitement rather than pretension.

That day became about far more than wine. It was about connection. We laughed easily, swapped stories, and talked in a way that never happens inside a warehouse or during a shift handoff. Titles and roles disappeared. We were just people spending time together. By the end of the day, after more than a few generous pours, spirits were high, and thankfully one of his friends had volunteered to drive. The warmth of that experience had little to do with alcohol or scenery. It came from genuine camaraderie and the ease of shared time.

Looking back, moments like that defined my time in Portland more than the work itself ever could. The job mattered, and the challenges were real, but it was the people who shaped the experience. Everyone I worked with carried their own history, skills, and perspective, and being invited into those stories created a sense of community that made even the hardest days manageable. That winery tour was just one example, but it captured something essential. Connection changes how work feels. Those memories, built on friendship and shared experience, are some of the ones I value most from that chapter of my life.

The connections built outside the workplace did not exist in isolation from the job. They were shaped by it, sustained by shared schedules, and often tested by the same pressures that defined the workday. Those relationships offered perspective and support, but they also underscored how closely work and personal life had become intertwined. Balance was less about separation than about managing overlap. Understanding that reality mattered before returning fully to the operation, where the demands of the role continued to press inward, regardless of where the workday officially ended.

Evaluators Lacked Relevant Operational Experience

Eight months into my time in Portland, I found myself standing at the edge of what felt like a major career leap, an interview for a manager role. This was not just another promotion. It felt like validation for everything I had been grinding toward since stepping into leadership. At Amazon, advancement is not earned solely through performance on the floor. It requires passing a POD, an internal interview gauntlet designed to evaluate candidates through the lens of Amazon's Leadership Principles. The process consisted of three back to back interviews, each forty five minutes long, conducted by a Senior Operations Manager, a mid level Operations Leader, and an HR representative. Together, they held complete authority over whether I would move forward.

The system was brutally binary. All three interviewers had to mark you as inclined. One dissenting vote meant failure, no matter how strong the other feedback might be. There were no second chances, no appeals, and no mechanism to contextualize a bad interview or a poorly framed question. Advancement did not hinge solely on competence or results. It hinged on perception. One misunderstanding, one poorly interpreted answer, or one interviewer having a bad read could erase months of effort in a single afternoon.

Every question was anchored to Amazon's Leadership Principles, and every answer had to follow the STAR format, Situation, Task, Action, Result. It was not enough to know how to do the job well. You had to narrate your success in a very specific way, complete with clean outcomes and measurable metrics. The pressure came not only from answering correctly, but from navigating the personalities on the other side of the table. Some interviewers, particularly those newer to leadership, approached the process less as an evaluation and more as a display of authority. The POD was not just a test of experience or judgment. It was a test of endurance, composure, and, frankly, luck.

During my POD, the cracks in Amazon's promotion system became impossible to ignore. The Senior Operations Manager, who was relatively new to the company, asked questions that had little relevance to the actual responsibilities of the role. It became clear very quickly that they did not fully understand the position they were evaluating me for. The second interviewer, a recently promoted leader, was more aligned but never pushed me to demonstrate my capabilities in a meaningful or challenging way. Then there was HR. While well intentioned, they lacked any real operational context. Their feedback felt detached, as if they were evaluating a role they had never seen in motion.

The most frustrating moment came when I was criticized for being too customer obsessed. I had shared a simple narrative I called "Sally Needs Her Dolly," a way to illustrate how workflow efficiency and on time delivery translated into real customer impact. It was designed to make logistics human, to connect metrics to outcomes that actually mattered. Instead of recognizing it as customer centric thinking, the panel dismissed it as being overly focused on the customer. The irony was hard to miss. Customer Obsession, the principal Amazon claims as its foundation, had somehow become a liability. At that point, it was obvious the decision had less to do with my performance and more to do with inexperience at the table and deference to the Senior Leader's opinion, which outweighed any meaningful discussion of readiness.

Despite weeks of preparation, much of it guided by my manager Aaron, I was ultimately marked as not inclined. Aaron had gone above and beyond to help me prepare. He built a detailed document of potential POD questions, walked me through difficult scenarios, and made sure I understood how to frame my experience effectively. When the decision came back, his frustration matched my own. He had watched my growth firsthand and believed I was ready. He even explored the possibility of an appeal, only to run into the same wall I did. The system allowed none. The decision was final.

I walked away from the process disillusioned. I had done what the system asked. I structured my responses correctly. I aligned them with the Leadership Principles. I demonstrated real experience and results. None of it mattered. My outcome had been shaped not by my capability, but by the subjectivity of the process and the uneven experience of those evaluating me. It was a sobering realization. At Amazon, advancement was not just about effort or skill. It was about navigating a system that often felt arbitrary, where inexperienced leaders could shape careers through decisions that lacked context, fairness, or insight.

As much as the experience stung, I refused to let it define me. I could see clearly that the rejection was not a reflection of my ability, but of a broken process. So, I did the only thing I could control. I went back to work. I doubled down on execution, on supporting my team, and on delivering results. At the same time, I began looking beyond Amazon's rigid promotion pipeline, exploring environments where experience and performance carried more weight than a checklist driven interview. I did not know what my next move would be, but I knew I would not allow a flawed system to determine my worth.

Ironically, later in my career, I found myself on the other side of the POD table as an interviewer. I carried that experience with me. For me, the process was never about posturing or rigid box checking. It was about understanding how someone thought, how they approached problems, and how they adapted under pressure. If a candidate struggled with a question, I did not penalize them. I turned it into a conversation. Sometimes those conversations taught me things about parts of Amazon I had never personally encountered.

In the end, the POD taught me lessons far more valuable than the promotion would have. It exposed the fragility of Amazon's leadership pipeline, the risks of elevating authority without experience, and the damage unchecked subjectivity can cause. More importantly, it clarified the kind of leader I wanted to be. One who led with fairness, empathy, and an awareness of the pressure people carry into those rooms. What initially felt like failure became a defining moment in my growth, reinforcing the kind of leadership Amazon often needed, but too rarely cultivated.

Being evaluated by less experienced leaders exposed a structural disconnect rather than an individual failing. Performance was often assessed through frameworks that valued adherence over understanding, and confidence over competence. Context was flattened, nuance lost, and judgment reduced to checklist compliance. What made the situation difficult was not disagreement, but the imbalance between authority and experience. Understanding how that gap shaped outcomes mattered before moving into environments where evaluations carried higher stakes and fewer opportunities for correction.

Leaving a Lasting Legacy

During my time at the station, I took real pride in leading initiatives that improved efficiency, strengthened driver engagement, and helped create a more positive work environment. I was never focused on metrics alone. I cared about building systems that made daily operations smoother while giving drivers a sense of ownership in their work. One of my most impactful contributions was the creation of detailed route staging maps. It was a simple concept, but highly effective. By giving drivers a clear visual layout of where their totes and packages were staged, the maps reduced the time spent searching, eased congestion, and improved overall safety in the load out area. The results were immediate. The chaos dropped, delays became less frequent, communication improved, and drivers walked in feeling prepared instead of frustrated. It was not just an efficiency win. It changed how the start of the day felt.

Another initiative, the PAC 10 mascot system, blended structure with personality and team spirit. Rather than relying on generic route identifiers that no one remembered, I assigned each driver a mascot tied to a PAC 10 collegiate team. These mascots became static markers for routes, making load out more intuitive while injecting some fun into the process. Drivers took pride in their assigned teams, talked trash in good humor, and leaned into the friendly competition that emerged. Engagement went up, confusion went down, and the floor felt more energized. It was a small change, but it mattered. Sometimes the difference between a process people tolerate and one they embrace is creativity paired with respect for the people doing the work.

Beyond the structural improvements, my highest priority was building real, personal connections with the drivers. Regular check ins were never just about performance metrics. They were opportunities to listen, identify problems early, and acknowledge the effort it took to get through difficult days. By staying visible and approachable, I made it clear that drivers could speak openly about challenges without fear of being dismissed. That openness encouraged initiative. Drivers began offering their own ideas for improving station flow and load out efficiency, which reinforced that their experience on the road mattered just as much as any spreadsheet.

To sustain performance without eroding morale, I introduced a Top 10 and Bottom 10 ranking system. The intent was never punitive. It was about clarity and growth. Top performers were recognized consistently, not just with numbers on a board but with direct acknowledgment of the work they were doing right. Those in the bottom group were not shamed or sidelined. They received focused coaching, clearer expectations, and practical support to help them improve. Over time, the system created accountability without resentment. Drivers understood where they stood, why it mattered, and what support was available to help them move forward. The result was stronger morale, better performance, and fewer recurring issues across the station.

At the core of my approach was a simple mindset. Drivers were my customers. My role was not just to enforce process or chase efficiency. It was to make sure drivers had the tools, structure, and support they needed to succeed. When drivers felt valued and backed by leadership, everything else improved. Engagement went up, operational friction went down, and collaboration replaced constant firefighting. Looking back, these efforts were never just about logistics. They were about building an environment where drivers felt invested, empowered, and motivated to perform at a high level, not because they were forced to, but because they wanted to.

Looking back, that period reinforced something I had come to believe deeply. Sustainable performance does not come from pressure alone. It comes from respect, clarity, and trust. The systems I built worked not because they were clever, but because they were designed with the drivers in mind. When people felt seen, supported, and included in the solution, they showed up differently. They cared more. They pushed harder.

They took ownership. In a company that often treated drivers as an interchangeable extension of the system, I tried to run my corner differently. Those efforts reminded me that leadership is not about control or optics. It is about creating conditions where people want to do good work. And when that happens, the results tend to take care of themselves.

Time to Transition

Both my manager and I knew it was time for me to return to Southern California, and the decision had nothing to do with dissatisfaction or unfinished business. It was simply the natural next step. Aaron was a dedicated family man with a wife and two young daughters, and he was also actively pursuing his long term goal of becoming an airline pilot. As his priorities and career trajectory began to shift, his availability at the station naturally narrowed. We both understood that change was coming, and neither of us viewed it as a loss. It was part of forward motion for both of us.

True to who he was, Aaron immediately went to work helping me line up what came next. He leaned on his network in Southern California and quickly connected me with a close friend of his who managed the station in Marwood Hills. That conversation turned into several more, followed by a knowledge test and a strong recommendation from Aaron. Not long after, I was offered a lateral role with a July start date. The opportunity felt right, but leaving Portland was harder than I expected. The relationships I had built there ran deep. While my coworkers were disappointed to see me go, their support never wavered, and that made the transition easier to accept.

The team made sure my departure did not go quietly. They organized a massive going away party at my wine buddy's house, and it turned into something far bigger than I ever anticipated. More than a hundred people showed up, including station leadership, Associates, DSP owners, and Delivery Associates. The night was full of laughter, stories, and an impressive amount of wine. The sense of community was overwhelming in the best possible way. Once again, Diana stepped in as my designated driver and made sure I got home safely, saving me from myself. The next morning, however, was far less forgiving. The headache I woke up with felt like punishment for every glass I should have skipped.

I had just four days to pack up and prepare for the move back south. To make things more complicated, wildfires had erupted across southern Oregon, filling the air with smoke and turning parts of the drive into a logistical challenge of its own. Visibility dropped, road conditions changed without warning, and patience wore thin. With the help of GPS and a series of back roads, I managed to navigate around the worst of it and make my way through safely. Before I knew it, Portland was behind me. I was heading back to Southern California, carrying a trunk full of belongings, a head full of memories, and a quiet readiness for whatever waited next in Marwood Hills.

As I crossed the California line, the weight of Portland slowly gave way to a steadier sense of resolve. That chapter had given me more than experience. It gave me confidence, relationships, and a clearer understanding of the kind of leader I wanted to be. I was not leaving anything behind unfinished. I was carrying it forward. The road south felt less like a departure and more like a continuation, a reminder that growth rarely announces itself with certainty, only with movement. Marwood Hills was waiting, and for the first time in a long while, I felt ready for whatever came next.

Settling into Marwood Hills

When I transferred to the Marwood Hills Distribution Center, I moved laterally into a Return to Station Shift Assistant role. The position placed me at the end of the delivery cycle, managing the flow of vans returning after their routes and dealing with the packages that had not reached customers.

My primary responsibility was making sure undelivered packages were accounted for, processed, and prepared for another delivery attempt. Depending on the time and available capacity, packages could be reassigned to a Delivery Service Provider, sent out with a Flex driver, or held for the next delivery cycle. The work required precision and quick judgment because every returned package represented a customer promise the station was still expected to keep.

Marwood Hills also introduced me to the Manager's Closet, a locked room stocked with candy bars, frozen pizza pockets, sodas, and other snacks. Most stations had some version of it, and during long shifts it became an unofficial survival station for managers who had little time to step away from the operation. The selection leaned heavily toward comfort food, but when the day stretched on, it was often exactly what we needed.

A designated manager was responsible for replenishing the stock, which typically involved a Costco run. Later in my tenure when my turn came, I saw an opportunity to introduce healthier options like bananas, tangerines, and low-calorie beverages. It was a small change, but it made a noticeable difference for those looking for lighter snacks amid the usual assortment of sugary and carb-heavy choices. This perk, though modest, felt like a thoughtful gesture from Amazon, emphasizing its support for operations managers navigating grueling schedules. Unfortunately, this benefit wasn't immune to change. When Andy Jassy stepped into the CEO role, the Manager's Closet underwent a noticeable scale-down. The variety dwindled, and the budget appeared to be trimmed, a disappointing shift for many of us who had come to appreciate this small yet meaningful resource. It was a subtle change, but it reflected the broader cultural shifts taking place within Amazon under its new leadership, marking the end of an era of small, morale-boosting gestures that had once made the grind a bit more bearable.

In my role, the RTS shift ran from 3 PM to the early hours of 1 or 2 AM, immersing me deeply into Amazon's complex logistics network. This position offered a unique vantage point, allowing me to witness the entire lifecycle of a package, from sorting to final delivery. It demanded constant engagement with various teams, including Sortation, Flex, and Dispatch, each presenting its own set of challenges and workflows. The fast-paced environment required seamless communication and operational precision to ensure timely deliveries, making the role both demanding and rewarding. Initially, I was paired with Andre, a seasoned Shift Assistant with a strong grasp of the processes but limited people skills. His "show once, then you're on your own" approach created a steep learning curve, forcing me to navigate the RTS process independently. While it was challenging at first, it pushed me to develop a deeper, more comprehensive understanding of the role. Over time, I grew to appreciate the shift for its hands-on engagement, both with associates and DSP owners, as I refined my skills in managing relationships and solving problems within this dynamic environment.

Settling into Marwood Hills changed the surroundings, but it did not immediately change the rhythm of my life. The move brought a sense of stability, yet familiar routines and work pressures followed, revealing how easily the same demands could reappear in a new setting. What changed most was my perspective: I began to see that geography offered only temporary distance, while the deeper patterns of work, pressure, and adjustment required something more fundamental to change.

Leadership and Amazon's Cultural Shift

This position proved invaluable in honing my DSP management abilities and offering insights into the operational and business pressures faced by DSP owners. I gained a deeper understanding of their daily struggles, from balancing labor schedules to meeting Amazon's stringent performance metrics. The experience sharpened my problem-solving skills and broadened my perspective on the interconnected challenges of logistics and business management. What truly captured my interest, however, was diving into the station's operational data. Leveraging my analytical background, I uncovered patterns and inconsistencies that helped streamline processes and improve performance. I also began creating Standard Operating Procedure (SOP) documents to guide new hires and ensure operational consistency, resources that are still in use today. This data-driven approach became one of my defining strengths, allowing me to make meaningful contributions to the station's ongoing success.

Ultimately, my time at the station became a period of real and lasting growth. I developed a strong command of RTS operations while sharpening my leadership instincts, analytical thinking, and ability to solve problems under pressure. Working through the complexities of the Flex program and the often-delicate relationships with DSPs gave me a clearer understanding of how Amazon's logistics ecosystem actually functioned in practice, not just on paper. I left that role with a stronger sense of where I could add value and how to do so effectively inside a system that demanded both precision and adaptability.

My new manager, Scott, represented a leadership style that was both refreshing and unconventional. He carried himself with an easy, laid-back demeanor, but underneath that calm was an exceptional depth of operational knowledge. Few people I met had a firmer grasp on how the station truly ran. What stood out most, though, was his fluency in Amazon's endless web of acronyms. He navigated that shorthand effortlessly, as if it were a second language. For me, it often felt overwhelming. Amazon leadership conversations could become so saturated with acronyms that meaning was lost entirely, leaving anyone outside that inner circle struggling to follow along. At times, it felt less like a tool for efficiency and more like a way to sound authoritative without actually saying anything substantive. Scott, however, was different. He knew the language, but he also knew the work, and that combination made his leadership both credible and effective.

Scott broke that mold completely. Despite his fluency in Amazon's internal language, he chose clarity over complexity every time. He rarely leaned on acronyms, preferring plain, direct language that actually conveyed meaning instead of obscuring it. That approach was transformative for me. Without having to mentally translate layers of jargon, I could focus on execution, strategy, and problem solving. It sharpened my ability to manage processes effectively and removed unnecessary friction from the work. Scott's transparency was not accidental. It reflected a leader who cared less about sounding impressive and more about making sure his team truly understood what needed to be done and why.

What made Scott truly unforgettable, though, was his background. Before Amazon, he had been a performer, singing and playing guitar in clubs and bars, bringing 90s hits to life night after night. That experience was not just an interesting detail from his past. It shaped how he showed up as a leader. His comfort connecting with people, reading a room, and keeping energy balanced came straight from that world. Music carried over into the workplace in subtle but meaningful ways, creating moments of levity and connection that were rare inside the pressure cooker of operations. Scott proved that leadership did not have to be rigid or impersonal to be effective. By combining clarity, competence, and a very human sense of joy, he created an environment where people could perform without feeling crushed by the weight of the system.

Scott had a habit of bringing his guitar to the station, turning late-night shifts into something extraordinary. Around 11 p.m., when the grind began to wear on everyone, he'd pick up his guitar and strum out familiar 90s songs, his voice so close to the original artists that it felt like a private concert. These impromptu performances transformed the atmosphere. During high-stress times like crash sorts, when the team was tasked with manually sorting massive backlogs of undelivered, returned, or missorted packages, Scott's music provided a much-needed respite. Crash sorts were grueling, requiring intense focus and speed, often leaning on the FLEX program to ensure packages reached their final destinations. Yet, in the midst of the chaos, Scott's music created a sense of camaraderie and lightness, reminding us that even the hardest moments could have their share of humanity.

Scott's influence reached far beyond moments of humor or levity. He became a mentor in the truest sense, guiding me through Amazon's complex operations with patience, clarity, and a genuine desire to see me succeed. Despite being nearly two decades younger than I was, he never treated the age difference as relevant. He listened to my perspective, respected my experience, and approached our working relationship as a partnership rather than a hierarchy. In an environment where leadership could often feel rigid or dismissive, that mutual respect was both rare and deeply empowering.

Looking back, Scott left a lasting mark on me, not just because of his operational strength, but because of the humanity he brought into a demanding role. He showed me that leadership is not only about driving results or mastering systems. It is about connection, trust, and the ability to make difficult challenges feel manageable rather than overwhelming. Through his mentorship, his authenticity, and even the quiet influence of his musical background, he modeled a kind of leadership that felt real and grounded. That example has stayed with me, shaping how I think about what it truly means to inspire a team and lead with both competence and heart.

Even now, I find myself listening to 90s bands like Third Eye Blind, Matchbox Twenty, Incubus, and The Offspring, all thanks to Scott. His ability to blend work with genuine camaraderie made long, punishing shifts feel lighter and more human. Scott did not simply manage tasks or schedules. He inspired people, mentored them, and created an environment where the work felt connected to something larger than metrics and throughput. It is a leadership style I still admire and consciously try to emulate.

Around that same time, though, I began to notice a troubling shift in Amazon's culture. A new wave of leadership was moving into place, many of them lacking a real understanding of last mile delivery and station level operations. They brought with them big ideas that sounded good in meetings but often collapsed under the weight of real-world execution. What made the shift especially concerning was how it collided with one of Amazon's stated Leadership Principles, "Have Backbone; Disagree and Commit." In theory, dissent was supposed to be valued. In practice, it became dangerous.

Managers who spoke up, questioned flawed logic, or pushed back on unrealistic directives were not met with discussion or respect. They were met with hostility. Disagreement carried consequences. Sometimes it showed up as subtle intimidation. Other times it took the form of overt bullying or coordinated efforts to sideline or remove people who refused to blindly comply. It was a stark departure from the culture I had come to respect, and it marked the beginning of a realization that principles on paper did not always survive contact with power, ego, and insecurity.

No one was immune to this shift, not even Scott. I watched him confront it directly, though not without consequences. One incident in particular still stands out. Station leadership instructed Scott to move pallets out of a docked linehaul trailer, fully aware that his Trailer Dock and Release certification had expired. That certification must be renewed annually and is a strict safety requirement for opening and closing fifty five foot linehaul trailers. The process itself is simple, completed step by step on a tablet, but the consequences of getting it wrong are absolute. Setting foot inside a trailer without valid certification results in immediate termination, no exceptions. Scott refused, as he should have, knowing exactly what was at stake. The fact that leadership put him in that position at all spoke volumes about the environment that was taking shape.

That incident was not an outlier. Scott's integrity, critical thinking, and willingness to push back made him a target in a culture that increasingly valued blind compliance over competence. Leaders like Scott, who encouraged creativity, respected their teams, and understood operations deeply, were becoming rare. In their place came leaders who treated metrics as doctrine and followed protocol rigidly, even when it actively undermined efficiency, safety, or morale. The Amazon I had once admired, a place that rewarded bold thinking and practical problem solving, was slowly being eclipsed by a culture driven by fear, conformity, and unchecked ego.

Eventually, Scott reached his limit. After fighting constant battles with a leadership group more focused on control than collaboration, he chose to walk away. He left to run operations at a warehouse for a well known shoe manufacturer, a role that offered him the autonomy and respect he had been denied. His departure was deeply personal for those of us who had worked closely with him, but it was also symbolic. It was another clear signal that something fundamental was breaking. Talented, capable leaders were not being pushed forward. They were being pushed out.

That exodus has never really stopped. Amazon has continued to lose some of its strongest leaders, not because they lacked results, but because they refused to operate in an environment that prioritized numbers over people, fear over innovation, and intimidation over collaboration. Watching that transformation unfold has been both frustrating and deeply disappointing. The company that once inspired me, one that claimed to value backbone and rewarded principled disagreement, is slowly fading. In its place stands a culture that feels smaller, harsher, and far less capable of sustaining the very ingenuity it once celebrated.

Looking back, that stretch of my career marked a quiet but irreversible shift in how I saw Amazon. The work was still complex, the scale still impressive, and the problems still worth solving, but the culture that once made those challenges energizing was eroding fast. Watching leaders like Scott leave did more than disappoint me. It clarified things. When a system consistently sheds its most capable, principled people, it is not evolving. It is protecting something brittle. That realization stayed with me. It reshaped how I understood power, leadership, and loyalty inside large organizations. I was still learning, still contributing, still growing, but for the first time, I could see the cracks clearly enough to know they were not accidental. They were structural. And once you see that, you cannot unsee it.

Rising to Leadership and Inventing

After Scott left, I stepped into the role of shift leader, overseeing station operations from 3 p.m. to 2 a.m. It was both an opportunity and a trial by fire. Once senior leadership walked out around 6 p.m., the entire station landed on my shoulders. This was a billion-dollar operation, and for the rest of the night, I was solely responsible for everything that happened inside those walls. The weight of that responsibility hit fast and never really eased up. Things went sideways more often than anyone would admit, and I constantly found myself walking a fine line between being authoritative enough to keep control and being pushed into decisions that felt uncomfortably close to dictatorial just to keep the building from unraveling.

What became clear just as quickly was that I was not completely on my own. One of the station leaders who had originally brought me over from Portland became an anchor during that transition. Even in a high-pressure environment that rarely slowed down, he made time to walk me through the realities of station management, especially when it came to labor decisions and package status changes. His support went well beyond technical guidance. During moments when I second-guessed myself, he offered steady encouragement and direct, honest feedback. That mentorship gave me the confidence to own the role fully and move through each shift with clearer judgment and a stronger sense of purpose.

While the role required a high degree of independent decision making, it did not take long for me to recognize that real success depended on building a strong and cohesive team. That realization pushed me to invest heavily in training my RTS associates, a choice that ended up transforming both the team and the station. I began delegating specific process driven responsibilities, giving associates clear ownership over their work instead of treating them as extra hands. They responded immediately. Not only did they meet expectations, they consistently exceeded them, proving that capability had never been the issue. Trust had been. As that trust took hold, so did a stronger sense of accountability and pride across the team.

With the core RTS processes running smoothly, I was able to redirect my energy toward deeper engagement with DSP owners and Delivery Associates. Building those relationships became central to how I led. Spending time with the people actually responsible for last mile execution gave me insight no dashboard ever could. I heard frustrations early, understood constraints more clearly, and could address issues before they escalated into failures. Those conversations reshaped how I viewed the operation. It was no longer about pushing volume through a system. It was about supporting the people who carried the weight of that system every day.

That shift brought clarity. Logistics was not just a network of packages and metrics. It was a web of human relationships that required trust, communication, and shared problem solving to function well. This period marked meaningful growth for me, both personally and professionally. With the guidance of a strong mentor and a team willing to rise to the challenge, I began to define a leadership style grounded in empowerment, adaptability, and collaboration. It was a style shaped not by authority, but by respect, and it became the foundation for how I would lead moving forward.

At the time, our station operated with ten Delivery Service Partners (DSP), which expanded to include five additional seasonal DSPs during peak periods. Managing this dynamic system gave me deep insights into DSP operations and the challenges faced by drivers, from navigating road conditions to adhering to Amazon's stringent policies and procedures. These firsthand experiences also revealed significant operational gaps, particularly the lack of visibility into driver performance. For instance, when drivers returned with undelivered packages, there was no historical data to determine if this was an isolated incident or part of a recurring pattern. This lack of insight created inefficiencies, preventing targeted interventions to address persistent issues.

Recognizing this problem, I decided to tackle it head-on by developing a software program called "DSPTrek." This tool interfaced with several of Amazon's existing systems to compile and analyze data in ways that had not been done before. I drew data from the real-time routing system, which tracked package deliveries by route, which linked route IDs to specific drivers. These two systems provided the foundational data, enabling DSPTrek to connect delivery metrics directly to individual drivers and routes.

With this framework, DSPTrek captured detailed metrics, including the number of packages assigned to a route, how many were delivered successfully, and the overall delivery percentage. By running the program multiple times daily, I could provide up-to-date performance snapshots for each driver and generate historical views over several weeks. Drawing on my previous experiences in Portland, I included a feature that highlighted the station's top and bottom driver performers each day. These insights gave DSP owners a clear picture of who was excelling and who might need additional support or coaching.

The impact of DSPTrek was immediate and significant. Armed with actionable, data-driven insights, DSP owners began addressing underperformance with targeted coaching while simultaneously rewarding their top-performing drivers. This balanced approach improved morale, encouraged accountability, and drove measurable results. One of the most notable outcomes was a steady increase in our station's Delivery Completion Rate (DCR), a key performance metric. DSPTrek's ability to identify trends and outliers transformed how we approached driver performance management.

Years later, Amazon introduced its own application, which provided similar functionality to DSPTrek. While it was gratifying to see the broader adoption of a concept I had pioneered, the lack of acknowledgment from higher leadership was a familiar sting. Despite the program's success and the noticeable improvements, it brought to our station's operations, my contributions went unrecognized. By then, I had come to accept that Amazon's culture often prioritized results over individual contributions, even when those contributions drove meaningful change.

Nevertheless, the experience reinforced my belief in the power of innovation and data-driven solutions. DSPTrek wasn't just about improving metrics; it was about empowering DSP owners and drivers with the tools to succeed, creating a culture of accountability and recognition that extended beyond numbers. Even without formal acknowledgment, the satisfaction of knowing that DSPTrek made a tangible difference kept me motivated. It served as a reminder that leadership isn't always about external validation, it's about finding solutions, supporting your team, and driving improvements where they matter most.

By the end of that chapter of my time at Amazon, I understood something that would follow me for the rest of my career there. The system would gladly absorb good ideas, real effort, and measurable results without ever absorbing responsibility for the people behind them. I had proven to myself that I could run a building, earn trust, build teams, and create tools that made the operation better in ways that actually mattered. But I had also learned that recognition was optional, protection was conditional, and sustainability was never part of the design. The work was real. The impact was real. And so was the quiet understanding that doing the right thing inside Amazon did not guarantee safety, credit, or longevity. It only guaranteed that more weight would eventually be placed on your shoulders.

Unlikely Lessons from Delivery Service Providers

Working closely with Delivery Service Provider (DSP) owners and their teams became one of the parts of the job I valued most. Each DSP brought a unique perspective, shaped by the owners' diverse professional backgrounds and entrepreneurial ventures. These small-business owners adapted their hard-earned expertise to Amazon's high-pressure Last Mile Delivery model, often in creative and inspiring ways. I genuinely enjoyed engaging with them, learning from their experiences, and witnessing the value they added to an otherwise rigid system. Their determination to succeed was evident in how they motivated their teams, overcame daily challenges, and maintained a relentless focus on delivering excellence on the road.

Through those collaborations, I gained a far deeper understanding of the delivery operation from the DSP owner's point of view. I saw how quickly pressures upstream could cascade down to the Delivery Associates. Staffing shortages, morale issues, and Amazon's constantly escalating expectations pushed owners to their limits. Yet many of them showed an impressive level of commitment to their teams. I watched owners personally step in to cover routes, invest extra time in training, and find creative ways to keep drivers motivated during periods of sustained stress. Balancing empathy for their people with the unforgiving demands of the business required a level of resilience that rarely showed up in metrics or reports.

One DSP in particular left a lasting impression on me. The company was called 250, and they stood out immediately. Many of the drivers were heavily tattooed, some from arms to necks, with a few discreetly covering teardrop tattoos beneath their right eyes. I would be lying if I said I was not cautious at first. This was my first time working with them, and they were a seasonal DSP, which only added to my uncertainty. That initial impression faded quickly. Beneath the rough exterior was one of the most disciplined and tightly run teams I encountered.

Unlike the ongoing drama and inconsistency that plagued some other DSPs, 250 operated with a level of structure that felt almost military. Roles were clear. Expectations were enforced. The team moved with purpose and cohesion. That discipline became especially evident during the holiday Peak season, when volume surged and Shipments Per Route jumped from an average of around 325 to well over 420 packages. Under that kind of pressure, cracks usually appeared. With 250, they did not. They held together, executed cleanly, and proved that strong leadership and unity could outperform appearance, assumptions, and even the chaos of Peak.

This period often tested the limits of DSPs. Many DAs balked at the increased workload, claiming it was impossible to complete the routes within the allotted time. For me, managing RTS during these high-stakes moments was particularly challenging. My software program, DSPTrek, became indispensable, helping me identify underperforming DAs who frequently returned with undelivered packages. By analyzing delivery patterns, I identified drivers who were taking extended breaks or showing a lack of urgency and pushed them back out to complete their routes. This strategy wasn't always well-received, and resistance was common, requiring me to collaborate closely with DSP owners to enforce compliance.

One memorable evening, several DAs from 250 returned to the station early with undelivered packages. When I asked them to head back out to complete their routes, they flatly refused, and one even threatened me, escalating an already tense situation. I immediately reported the incident to Jesus, 250's owner. His response was swift and decisive. Within minutes, every driver returned to the road to complete their deliveries. Later that evening, the same DA who had threatened me approached with a sincere apology, promising it wouldn't happen again. I couldn't help but notice fresh cuts and bruises on his face that weren't there earlier.

Jesus approached me later that night and made his expectations unmistakably clear. He told me he valued operational excellence and had no tolerance for incompetence. There was a smirk when he talked about money, about making a lot of it, and about how seriously he took performance. He spoke calmly, confidently, and with an edge that left no doubt about who was in charge. For a brief moment, a thought crossed my mind about whether 250 might have ties to something more organized than a typical delivery business. I never had proof, and I never pursued the idea. What mattered was what I could see. Their discipline, loyalty, and results were real.

From that point forward, 250 became one of the station's strongest performing DSPs. They consistently hit a one hundred percent Delivery Completion Rate with zero customer complaints. Week after week, they delivered clean execution with no noise and no excuses. Despite their intimidating appearance, they proved to be a critical asset to the operation. Over time, my initial caution turned into genuine respect.

Their approach was not polished or conventional, but it was effective. They ran their business with clarity, accountability, and an unwavering focus on results. Working with them reinforced something I had already begun to learn. Leadership and success do not always look the way you expect them to. Sometimes they show up in rough packaging, speak plainly, and deliver outcomes that everyone else struggles to achieve.

Working with 250 crystallized a lesson that stayed with me long after that season ended. Amazon liked to pretend it could standardize leadership, culture, and performance into neat boxes that looked good on paper, but the real world never complied. Results did not come from appearances, polished language, or corporate alignment. They came from clarity, consequences, and leaders who meant what they said.

Watching 250 succeed forced me to confront how often Amazon misunderstood where real strength lived. The system demanded obedience, but what actually carried it through Peak was discipline, trust, and accountability enforced at the human level. That gap between what Amazon preached and what actually worked would only widen over time, and this was one of the first moments where I saw it clearly, without illusion.

Manager's Balancing Act

At Amazon, the workload placed on managers is overwhelming, a reality well understood by anyone who has held the role. From the first day, managers are handed a list of responsibilities large enough to consume an entire ten hour shift, and that is before anything unexpected enters the picture. Leadership regularly layers on last minute tasks and new priorities, stretching days well beyond what is officially scheduled. While managers are technically assigned four ten hour shifts each week, complete with breaks and a lunch period, the reality rarely matches the policy. Most managers work straight through, skipping breaks and pushing past fatigue just to keep up with the volume of demands placed on them.

The idea of working smarter, not harder, often feels disconnected from the actual conditions of the job. The workload is unpredictable by design. Sudden volume spikes, surprise audits, and unplanned assignments routinely derail even the most carefully structured schedules. Despite that volatility, expectations remain fixed. Performance targets do not move, and flexibility is minimal. Managers are expected to adapt instantly, juggle competing priorities, and solve problems in real time, all while maintaining flawless execution.

RTS operations are a perfect example of how fragile that balance can be. The process relies on multiple interdependent steps, drivers returning on time, associates sorting accurately, and upstream routing being completed correctly by the previous shift. When any one of those elements breaks down, the impact ripples through the entire operation. Delays compound quickly, yet accountability never shifts. Even when the root cause sits completely outside their control, managers are still held responsible for hitting deadlines and maintaining performance. It is a constant exercise in pressure management, one where effort and endurance often matter just as much as skill.

This relentless workload often crosses into unsustainable territory. While Amazon's emphasis on efficiency and innovation sounds forward-thinking on the surface, in practice, it frequently feels like a justification for piling more onto managers' already overflowing plates. The company's approach fails to account for the countless interdependencies inherent in the work, where tasks are delayed not by inefficiency but by factors entirely outside a manager's control. These challenges test the limits of even the most skilled and experienced managers, demanding sacrifices in personal well-being and work-life balance that many find difficult to endure.

Ultimately, the role of a manager at Amazon is as rewarding as it is draining. The opportunity to lead and innovate within one of the world's most advanced logistics networks is significant, but the unrelenting pressure often overshadows the potential for fulfillment. No matter how dedicated or capable a manager may be, the constant demands, lack of flexibility, and culture of relentless output can push them beyond sustainable boundaries, leaving them overextended and, at times, undervalued. It's a role that highlights both the remarkable ambition of Amazon's operational goals and the profound cost of achieving them.

At the Marwood Hills station, parking challenges were a persistent issue that highlighted the logistical complexities of managing a high-volume operation. The lack of on-site parking for DSP vans and DA vehicles forced Amazon to contract with offsite businesses, utilizing their lots to accommodate the overflow. While this solution was practical in theory, it brought with it a host of new headaches. I was tasked with managing these offsite lots in addition to my already full slate of responsibilities. This included creating and maintaining organized parking lanes, assigning spaces for DSPs, and fielding constant complaints from DSP owners unhappy with their lot assignments. It felt like an endless cycle of adjustments and negotiations, with no perfect solution in sight.

The task was not assigned to me by accident. It grew out of escalating frustration from parking lot owners who were fed up with recurring issues tied to station operations. Complaints were constant. Trash left behind after load out, reckless driving, and even occasional physical confrontations between Delivery Associates had pushed some of them to the brink. For station leadership, many of whom were younger and lacked the life experience needed to navigate tense, real world conflicts, the situation felt unmanageable. The lot owners, on the other hand, were closer to my age, and leadership believed I stood a better chance of building rapport and calming the situation. They were right, but that did not make the responsibility any easier.

Managing those relationships required a careful balance of diplomacy and problem solving. The lot owners were justified in their frustration, and addressing their concerns meant dealing with everything from organizing regular cleanups to enforcing stricter expectations around driver behavior. At the same time, I had to account for the realities DSPs were facing, including limited parking options and constant time pressure during load out. Both sides felt unheard, and both expected solutions. There were no clean wins, only compromises that required patience and persistence.

It was a thankless role, but an important one. It exposed a broader truth about Amazon's operation. The consequences of logistical decisions rarely stop at the station doors. They spill into surrounding neighborhoods, businesses, and communities. Solving those problems requires more than process knowledge. It demands empathy, resilience, and a willingness to engage with stakeholders far removed from the warehouse floor.

One particularly vocal lot owner, Roger, left a lasting impression on me. He owned a twenty acre dirt field that had once been a cow pasture. The land had been leveled and covered with gravel to keep dust down and prevent vans from sinking into mud during the rainy season. Beyond that, the property had little practical use, but Amazon paid him well to lease it as overflow parking for DSP vans. On the surface, it seemed like a straightforward arrangement, and at first, I assumed there was little reason for dissatisfaction. That assumption did not hold up for long. Roger was deeply frustrated by trash left behind, reckless driving through his property, and, more than anything else, the lack of communication from station leadership. Calls went unanswered. Concerns were brushed aside. Over time, that neglect became the core of his anger.

What surprised me was how quickly our relationship shifted once someone simply took him seriously. I made a conscious effort to address his concerns directly, often spending extra time outside my normal shifts to follow up and resolve issues. I did not dismiss his complaints or treat them as inconveniences. I listened, explained what I could control, and followed through when I said I would. That alone changed everything. What began as a stream of frustration slowly turned into a working relationship built on mutual respect. Before long, Roger saw me as his primary point of contact with Amazon. He called me regularly, sometimes daily. It added to an already heavy workload, but I understood why it mattered to him, and I did my best to be consistent and responsive.

That kind of dynamic extended beyond Roger. I built similar relationships with other lot owners, easing tensions and stabilizing partnerships that had been on the verge of falling apart. Roger's case, however, stayed with me. About a year after we began working together, I received an unexpected call from his wife. She told me that Roger had passed away. During the conversation, she mentioned that he often referred to me as "the voice of Amazon." Hearing that stopped me cold. I had never thought of myself that way, but knowing that my efforts had meant something to him, especially in a situation that caused him so much frustration, was deeply moving. Her words stayed with me long after we hung up.

Roger's story became a quiet reminder of why the human side of the job mattered so much. Managing offsite parking was one of the most thankless parts of my role, full of complaints, tension, and constant adjustments. There were no metrics that captured that work, and no recognition for doing it well. Yet moments like that made the extra effort worth it. They reinforced something I came to believe strongly. Beneath the spreadsheets, performance targets, and logistics, it is empathy, respect, and consistency that leave the most lasting impact.

By the time I stepped back and looked at everything layered onto that role, the pattern was impossible to ignore. Amazon measured managers almost exclusively by output, but the work that actually kept the operation stable lived in the margins. It was the conversations that defused conflict, the follow-through no one tracked, and the relationships built with people who were never listed on an org chart. Those efforts absorbed time, energy, and emotional weight, yet they never appeared in a metric or a performance review. They were simply expected, silently folded into an already impossible workload.

That realization lingered with me. The system rewarded scale and speed, but it relied on individual managers to absorb the human cost without acknowledgment or relief. Being "the voice of Amazon" was not a title anyone assigned, and it certainly was not one the company recognized. It was something earned by showing up, listening, and caring when it would have been easier not to. Over time, moments like that clarified what kind of leader I wanted to be and what kind of system I was operating inside. One valued numbers above all else. The other valued people. Holding both at once was exhausting, and it would not be sustainable forever.

The Strain on Associates and the Push for Change

One of the most glaring issues I observed during my time at Amazon was the lack of empathy displayed by station leadership in their treatment of associates. Too often, associates were reduced to numbers on a spreadsheet, their individuality and humanity overshadowed by a system laser-focused on efficiency and metrics. This dehumanizing approach created a sense of alienation among the workforce, fueling dissatisfaction and opening the door for unionization efforts. For many associates, unions represent a hope for collective bargaining and a voice in a system where they often feel unheard.

One of the clearest illustrations of this tension played out at a large station in the Southeast. When employees there began pushing for unionization, Amazon's response was swift and uncompromising. Rather than engage or negotiate, the company shut the station down entirely, disbanding the operation and eliminating more than 2,500 jobs overnight. The message was unmistakable. Organizing would not be tolerated, and the consequences would extend far beyond the individuals involved.

A similar pattern surfaced in Southern California. When Delivery Associates at a DSP successfully unionized, Amazon terminated the DSP's contract outright. The fallout was immediate and severe. Roughly fifty drivers lost their jobs, and the DSP owner saw their business collapse in an instant. Years of investment, risk, and effort were wiped out with a single decision. It is difficult to imagine the financial and personal impact of that outcome for everyone involved.

Together, these incidents revealed how willing Amazon was to use its scale and leverage to suppress unionization. The power imbalance was stark. Workers and small business owners were left with little recourse, while the company absorbed the disruption without meaningful consequence.

Watching these situations unfold reinforced a troubling reality. When control is prioritized above dialogue, the human cost becomes collateral damage, quietly absorbed by those with the least ability to protect themselves.

Beyond the more drastic measures, Amazon's stance on unionization showed up in quieter but telling ways. Managers across the network were required to sit through regular training sessions on how to handle union conversations and respond to media inquiries. It was hard not to read that for what it was: a defensive posture that revealed just how uneasy leadership was about the growing push for worker representation. But the real issue was never unions. It was how associates were treated, the very people who carried the weight of Amazon's operation every day.

Most associates worked grueling ten-hour shifts that included a thirty-minute unpaid lunch and two paid fifteen-minute breaks. In reality, many skipped their breaks altogether, trading their own well-being for the pressure to keep up with the workload. That choice came with a cost Amazon didn't ignore. When associates failed to clock out for lunch, the company ended up paying for the extra time. Leadership eventually caught on, and the response wasn't to slow the work or adjust expectations. Instead, Amazon rolled out a policy mandating that associates take their full thirty-minute lunch. Not complying meant write-ups and, in some cases, termination. It was another example of how control often took priority over addressing the conditions that pushed people to skip breaks in the first place.

These policies, while framed as protections for associates, often felt like just another layer of control layered onto an already oppressive system. For many employees, forced breaks did little to address the real problem. The issue was not a lack of mandated rest. It was an unrelenting workload paired with expectations that were simply unrealistic. When the volume never slowed and staffing never truly caught up, a scheduled break felt more symbolic than restorative.

That disconnect between leadership intent and frontline reality only deepened existing frustrations. Associates could see that the surface level fixes did nothing to relieve the pressure they faced every day. Instead of listening to what people were actually asking for, leadership responded with policies that looked good on paper but failed in practice. That gap in understanding continues to fuel the very grievances that drive unionization efforts in the first place.

While the long term outcome of Amazon's ongoing conflict with labor organizations remains uncertain, one truth is hard to ignore. The systemic disregard for associate wellbeing sits at the center of the issue. Until that reality is confronted honestly and meaningfully, trust will continue to erode. And without trust, Amazon risks losing not just goodwill, but the loyalty of the people who keep its logistics network running day after day.

Low associate morale has long been a challenge across Amazon's network, and Marwood Hills was no exception. This issue became especially pronounced with the launch of the Same Day Sort (PFSD) program, which aimed to handle packages ordered early in the day and guaranteed for same-day delivery. Forming the PFSD team required pulling associates from various shifts, leaving noticeable gaps in enthusiasm and energy. During Stand-Up Meetings (SUMs), the lack of motivation was palpable. This disengagement frequently translated into errors in the sorting process and incomplete package routing, undermining the efficiency of the operation.

Recognizing the need for change, I set out to revamp the SUMs, transforming them from routine briefings into something more engaging, interactive, and inspiring. Instead of solely focusing on safety and work standards, I expanded the scope to include humor, knowledge-sharing, and a sense of collaboration. One of the most impactful changes was introducing a "Fun Facts" section on the Gemba board. These facts covered a range of topics, from explaining Amazon-specific acronyms like UTR, GRIP, MILK, DSP, and DA, to insights about the Package Life Cycle and highlights of Amazon's innovations, such as the Blue Origin space program, AWS, and Audible. I even delved into broader tech topics, discussing the history of cloud computing, the iPhone, and Steve Jobs' vision for the future. These sessions sparked curiosity among associates, providing them with a deeper understanding of Amazon's initiatives and a new appreciation for the technology shaping their work.

To help shift the tone on the floor, I started opening each SUM with a bit of humor. I would tell the team, "You're Santa Claus today, so smile like you just got free cookies, have fun like the reindeer are watching, and be safe because nobody wants a sleigh pile up." It was simple and a little corny, but it worked. The lighthearted approach broke through the fatigue and reminded associates that their work mattered, not just to the station, but to a global operation that depended on them showing up engaged and focused.

As morale lifted, engagement followed. Associates stopped passively listening and started contributing. Conversations during SUMs became more interactive, with people offering ideas instead of just waiting for instructions. One of the most impactful suggestions came from the floor itself. Someone proposed better use of the Big Iron, a multi story conveyor system designed for heavy duty sorting. At the time, we were leaning too heavily on manual crash sorts, which had become a serious bottleneck as package volume surged under the PFSD program. Shifting more volume onto the Big Iron changed everything. It reduced manual handling, cut down errors, and helped ensure routes went out more complete and on time. It was a clear reminder that when associates feel heard and respected, the best solutions often come from the people closest to the work.

Over time, those efforts reshaped the SUMs into something far more meaningful than a required briefing. They became moments of connection, collaboration, and shared growth. Associates began to feel genuinely valued and more closely aligned with Amazon's broader mission. Their ideas were not just acknowledged in passing. They were acted on, which reinforced a sense of ownership and pride in the work they were doing every day.

Watching that shift unfold was deeply rewarding. A team that had once felt distant and disengaged slowly became cohesive and invested, driven by shared goals rather than obligation. The cultural change at the station was not the result of sweeping policy changes or complex programs. It came from small, intentional adjustments. A bit of humor. A willingness to listen. An openness to learning from the people doing the work. It was a clear reminder that when people feel respected, heard, and connected to something larger than themselves, performance and morale tend to rise together.

That experience reinforced a truth that sat in direct contrast to how Amazon preferred to operate. Morale was not a mystery problem, and engagement was not something that had to be engineered through slogans or mandates. It responded almost immediately to being treated seriously, spoken to honestly, and included in the work rather than managed around it. The same associates who were written off as disengaged or unmotivated proved capable of insight, creativity, and ownership the moment someone bothered to create space for it. Nothing about the work itself had changed. The volume was still high. The pressure was still real. What changed was how people were seen.

That contrast became harder to ignore as time went on. On one side was a corporate posture built on control, discipline, and fear of dissent. On the other was a simple, human approach that required no policy rollout, no training deck, and no executive buy-in. Just presence, respect, and follow-through. Watching those two realities exist side by side made it clear which one actually worked. It also made clear how fragile that kind of progress was inside a system that did not value it.

In the end, the lesson was not about unions, metrics, or programs like PFSD. It was about empathy as an operational advantage that Amazon consistently failed to recognize. When people are treated like replaceable inputs, they eventually push back or shut down. When they are treated like partners in the work, they rise. That gap between what leadership feared and what the floor actually needed would continue to widen. And for those of us paying attention, it became impossible to unsee where the real problem lived.

The Complexity of Operational Processes

In Amazon's earlier days, the MILK Run process was a meticulous operation designed to correct routing errors for pallets of packages mistakenly delivered to the wrong station. These shipments were often destined for other Distribution Centers or Fulfillment Centers and had to be rerouted accurately before continuing on to the customer. The MILK Run was not simply about fixing mistakes. It was a labor intensive workflow that demanded constant attention to detail, with little margin for error, because every misstep risked compounding delays across the network.

This became one of the most data heavy and complex processes I encountered during my time with Amazon Logistics. Stations routinely received pallets and racks of misrouted packages. Some were labeled for entirely different stations. Others were inexplicably marked as delivered. A few were tied to the wrong facilities altogether. Untangling those errors was made even harder by system limitations. Tools like Carrier Operations Management Platform (COMP), which once allowed more flexible status corrections, were eventually replaced by Station Command, a platform that stripped away functionality that had made the process manageable. Each change added friction. Successfully navigating the MILK Run required patience, creativity, and a constant willingness to adapt as tools, rules, and expectations shifted beneath your feet.

At the center of it all was the TBA, Amazon's unique tracking barcode printed on every package. Correcting errors meant diving deep into each TBA's status, often requiring multiple adjustments in the routing system to bring the package to a "good routable" status. Some packages needed dozens of updates before their delivery path was fully corrected. With strict deadlines to reroute packages and constant pressure to minimize delays, the stakes couldn't have been higher. It was a test of both accuracy and speed, demanding relentless focus from everyone involved.

When I first encountered the MILK Run at the Marwood Hills station, I was struck by its intricacy. It took me over a month of dedicated effort to fully grasp every step and nuance of the process. Recognizing the challenges, I set out to create structure and consistency by developing a comprehensive Standard Operating Procedure (SOP). This detailed guide broke down the MILK Run into clear, actionable steps, providing a roadmap to navigate its complexity with accuracy and efficiency.

Once the SOP was finalized, I shifted my focus to training the 15+ associates on the RTS team. My goal was to empower them to handle the MILK Run independently and confidently. Through hands-on sessions, I taught them how to analyze TBA data, execute status updates, and ensure packages were properly rerouted. By the time I left Marwood Hills, every associate was proficient in the process. What had once been an overwhelming and error-prone task became a streamlined operation that associates could manage with precision, even under tight deadlines.

The impact of this effort extended beyond just improving efficiency. It fostered a sense of ownership and confidence among the team, who took pride in mastering such a challenging process. They no longer felt daunted by the MILK Run; instead, they approached it as a well-understood, manageable part of their workflow. Reflecting on this initiative, it became clear that the combination of clear documentation and thorough training was key to transforming what had been a daunting task into an integral and seamless part of station operations.

Looking back, the MILK Run became another quiet example of how much institutional knowledge Amazon was willing to risk losing at any given moment. The process worked not because the tools were well designed, but because people filled in the gaps with experience, workarounds, and persistence. When that knowledge lived only in someone's head, the system functioned just barely enough to survive.

When it was documented, taught, and shared, it became resilient. That distinction mattered far more than leadership ever seemed to acknowledge.

What stayed with me was how easily that progress could have disappeared. One transfer, one resignation, one burnout exit, and the entire process would have slipped back into chaos. There was no formal mechanism to protect or preserve the work that made the operation actually function. It relied on individuals caring enough to do more than was required, without any guarantee that effort would be valued or even remembered. The MILK Run taught me that excellence at Amazon often existed in spite of the system, not because of it. And it reinforced a growing realization that the company's greatest vulnerability was not volume or complexity, but its willingness to treat hard-won human expertise as expendable.

Empathy in the Workplace

I have often felt that empathy is the quality Amazon's work culture lacks most, yet it has always been central to how I lead. It is one of my greatest strengths, and I suspect it was shaped long before Amazon, through my experience as a father. Learning how to support, guide, and truly understand my children taught me patience and perspective, lessons that carried directly into how I viewed the people I worked with. Associates, in particular, bore the heaviest load while receiving the least support, and I made a conscious effort to see the person behind the job title. One of those people was Betty, who joined our RTS team a couple of months after I started and quietly left a lasting impression on me.

Like every new hire, Betty was paired with an ambassador to learn the core RTS processes. She was trained on Debrief, the MILK Run, Google Docs scanning, and managing Trouble Tickets. From the beginning, she stood out, though not for the usual reasons. It was not her speed or technical ability that caught my attention, but her quiet, introverted presence. She kept to herself, rarely initiating conversation, yet she was friendly enough and seemed to get along with her peers. A few weeks in, I began to notice that her performance in Debrief was falling behind others. Comments started to surface, subtle and indirect, from associates who hinted that she was not carrying her share of the workload, even though no one openly criticized her.

Rather than allowing the situation to escalate or leaving her to struggle in silence, I decided to address it directly yet compassionately. After a shift, I pulled Betty aside for a private conversation. I approached the discussion carefully, not as a performance review but as an opportunity to understand her perspective. Betty was clearly uncomfortable, but after some encouragement, she opened up. She confided that she had a learning difficulty, one that made it challenging for her to grasp new and complex processes, particularly technical ones. She admitted she was afraid of being labeled incompetent and felt embarrassed to seek help through HR programs, fearing it would only make her stand out more.

Her honesty stayed with me. I reassured her that she was not alone and that struggling did not mean she was failing. I offered to work with her one on one, outside the pressure of a normal shift, if that felt safer and more manageable. She hesitated at first, clearly unsure whether she wanted the extra attention, but eventually agreed. Together, we mapped out a plan built around her pace, not the station's. We broke the Debrief process into smaller, digestible pieces, focusing on one component at a time until she felt steady. Once that foundation was in place, we moved on to Trouble Tickets and COMP, taking each step slowly and deliberately. For the MILK Run, which was the most complex of all, I built her a custom roadmap she could follow without feeling overwhelmed. I stayed late after shifts and even came in on my days off, walking her through each task while also explaining the larger context, how her work connected to the station's broader goals and why it mattered.

Over time, something shifted. Betty's confidence began to grow in visible ways. She stopped second guessing herself and started solving problems on her own. The workflows that once intimidated her became familiar, even routine. She began taking ownership of her role with a quiet assurance that had not been there before. Watching that transformation was deeply rewarding. It was not just about improved metrics or cleaner processes. It was about seeing her stand a little taller, speak with more certainty, and move through her work without fear. The hesitation that once defined her presence faded, replaced by determination and pride. In that change, I was reminded why empathy mattered so much, not as a soft skill, but as a practical force that could unlock potential when systems alone could not.

As her confidence grew, so did her engagement with the team. Betty started interacting more with her peers, contributing to discussions, and taking greater initiative in her work. The once quiet and reserved associate was now a key part of the group, offering ideas and stepping up to handle tasks with enthusiasm. Witnessing her find her footing and truly flourish was one of the most fulfilling experiences of my time at Marwood Hills. By the time I left, Betty had evolved into a top performer, a remarkable shift from where she had started. She was even being considered for an RTS ambassador role, a testament to how far she had come.

Betty's evolution wasn't just about her technical skills; it was about the person she had become. She was no longer timid or withdrawn but self-assured, proactive, and a leader in her own right. Her journey was a powerful reminder of the transformative impact that empathy, patience, and personalized support can have in a demanding and often impersonal environment. Watching her growth reaffirmed the importance of seeing the potential in people, meeting them where they are, and giving them the tools and encouragement, they need to thrive.

On my last day at Marwood Hills, Betty surprised me with a goodbye card and a dozen homemade cupcakes. It was a small, deeply personal gesture I hadn't expected at all. Her note hit harder than I was prepared for. She thanked me for my patience, for taking the time to explain things, and for making her feel capable in a role she once found intimidating. At the end, she wrote that I was the only one who had ever believed in her. That was a sentence I would hear again and again throughout my Amazon journey. It was one of those rare moments that makes you stop and take stock, a reminder that empathy matters, especially inside a fast-paced, high-pressure environment where people are often treated like numbers instead of humans.

Betty's story became a quiet but powerful reminder of what is possible when people are met with understanding, patience, and the right kind of support. She did not succeed because of some hidden, extraordinary talent. She succeeded because someone took the time to see her potential and gave her the space and guidance to grow into it. With that support, she exceeded every expectation and became a confident, capable member of the team.

For me, helping Betty was never just about improving performance or fulfilling a managerial obligation. It was about recognizing her worth as a person and making sure she felt seen, valued, and capable of succeeding on her own terms. Moments like that reinforced why empathy mattered so much to my leadership approach. When people feel supported instead of judged, they often rise far beyond what anyone initially expects of them.

Her transformation stands as a testament to the power of empathy to not only change individual lives but also to influence the work culture around them. Betty's success reinforced a truth that I carried with me long after my time at Amazon: when people are given the support they need to thrive, they can achieve extraordinary things, and those moments of connection can leave a lasting impact on everyone involved.

By the time I walked out of Marwood Hills for the last time, I understood that moments like Betty's would never show up in any report Amazon cared about. There was no metric for belief, no dashboard for patience, no KPI for staying late to help someone find their footing. Yet those moments were the ones that stayed with me. They carried more weight than any performance review or operational win because they revealed what leadership actually looked like when stripped of titles and expectations. It was quiet, personal, and often invisible to the system that benefited from it.

That disconnect never stopped bothering me. Amazon celebrated efficiency, speed, and scale, but it quietly depended on leaders willing to slow down when the system refused to. It relied on people who would absorb emotional labor without recognition, who would choose empathy even when it was inconvenient, and who would see associates as human beings in an environment designed to treat them as interchangeable parts. Betty did not succeed because Amazon made room for her. She succeeded in spite of a system that would have written her off without hesitation.

In the end, her note said more about Amazon than it did about me. If being seen and believed in felt that rare, that was the real failure. And it forced a question I could not ignore as my career progressed. If empathy was the thing that consistently unlocked trust, growth, and real performance, why was it treated as optional or even suspicious inside the culture? The answer would become clearer over time. But that moment, holding a handwritten card and a box of cupcakes, marked one of the first times I understood just how far apart Amazon's definition of leadership was from my own.

A Bittersweet Departure

After about ten months at Marwood Hills, I knew it was time to look for opportunities closer to home. The commute had always been a grind, and I could feel myself ready for a new chapter. I reached out to my station leader mentor for guidance, hoping he might know of something in the area. As it turned out, the timing could not have been better. He told me about an opening at the Ashford Distribution Center, the very station where I had first started. A manager there, who also happened to be a close friend of his, was looking to fill a role. With my mentor's recommendation in hand, I reached out immediately. Within days, I had an offer. It was technically a lateral move, but it came with a clear path to promotion and a dramatically shorter commute, which made the decision easy.

Leaving Marwood Hills was harder than I expected. In less than a year, I had built real relationships with the people there, and saying goodbye carried some weight. One person in particular stood out, Sergio, the morning sort manager. We had worked closely to keep our shifts aligned, and his consistency and support made a real difference in how smoothly things ran. On my last day, he surprised me with a parting gift, a Costco sized bottle of bourbon and an equally oversized bottle of Coke. With his trademark charisma, he insisted we head out to the parking lot to toast my next role. We laughed, talked, and shared a moment that perfectly captured the camaraderie we had built. It meant more to me than he probably realized.

Not long after I left, I got news that caught me completely off guard. Sergio had been let go for being overly friendly with some of the female associates. It was difficult to reconcile that with the person I had worked alongside, someone who always seemed professional and well liked. Even now, I find myself wondering what became of him. I hope he managed to land on his feet. He was a strong colleague and someone I would not hesitate to work with again under different circumstances.

While my departure from Marwood Hills was emotional, I left knowing my team was prepared to handle any challenges that came their way. Over the months, I had poured significant time and energy into training them, ensuring they had the skills and confidence to succeed without me. Watching their growth had been one of the most fulfilling parts of my role, and I left feeling proud of what we had accomplished together.

As I headed toward Ashford, I felt the familiar pull of nostalgia mixed with genuine excitement. I was leaving behind people and experiences that had shaped me in ways I did not fully appreciate at the time, and that was not easy. At the same time, there was a quiet energy in knowing I was stepping into something new, something that would challenge me in different ways.

That move marked the beginning of another chapter in my Amazon journey. It carried its own set of lessons, pressures, and moments of growth, some expected and others completely unforeseen. Looking back, it was another reminder that each transition brought more than a change of location or role. It continued to shape how I thought about leadership, resilience, and the kind of professional, and person, I was becoming.

As I drove away from Marwood Hills for the last time, the mix of relief and loss sat heavier than I expected. The move made sense on paper, shorter commute, clearer trajectory, familiar territory, but Amazon rarely allowed clean transitions. You carried people with you, moments with you, and unfinished questions that never got answered. I was leaving behind a team I trusted, relationships forged under pressure, and proof that leadership grounded in empathy could actually work, even inside a system that resisted it.

What I did not realize then was how much Marwood Hills had sharpened my awareness. It stripped away any remaining illusion that success at Amazon followed a straight or fair line. Good leaders could disappear overnight. Strong teams could be undone by forces far above them. And progress, no matter how real, was always fragile. I was heading back to Ashford more capable, more confident, and far more clear-eyed than when I had first walked into an Amazon building years earlier.

That clarity would matter. The next chapter would not be about learning how the system worked. I already knew that. It would be about learning how long someone could stay inside it without compromising who they were. At the time, I felt ready for what came next. I did not yet understand how much that question would follow me, or how close it would eventually cut.

Bullying Became Normal

In my early days at Amazon, I was captivated by the company's culture. A dynamic environment brimming with energy, ambition, and a bold willingness to think outside the box. There was an almost rebellious emphasis on innovation, where risk-taking wasn't just encouraged; it was celebrated. Failure wasn't something to be avoided but rather embraced as a stepping stone to success, a philosophy that allowed Amazon to push boundaries and redefine industries. This culture of experimentation fostered a sense of empowerment among employees, encouraging them to dream big and take chances without fear of reprisal. It was clear that this fearlessness, championed by Jeff Bezos, had been instrumental in the company's meteoric rise.

Bezos' leadership style was rooted in an entrepreneurial mindset that permeated every level of the organization. Big, audacious ideas were not merely welcomed, they were expected. Initiatives like Prime Day, Alexa, and Amazon Go reflected a company willing to invest heavily in ideas that, on the surface, seemed risky or even unrealistic. Failure was not treated as something to be avoided at all costs. It was viewed as a source of insight, a necessary step toward breakthroughs that could redefine entire industries. That philosophy created a culture where creativity thrived and employees felt they were contributing to something larger than themselves, a company unafraid to challenge convention and reset expectations.

That culture began to shift in July 2021, when Jeff Bezos stepped down as CEO and Andy Jassy, the architect behind Amazon Web Services, assumed the role. Jassy's success in building AWS was undeniable, but the leadership philosophy that worked so well in cloud infrastructure brought a very different tone to the rest of the company. AWS thrived on precision, stability, and risk mitigation; traits essential in enterprise technology. That approach, however, contrasted sharply with the bold, entrepreneurial energy that had defined Amazon's earlier years.

Under Jassy's leadership, the cultural shift became increasingly visible. The once expansive tolerance for experimentation began to narrow. Risk taking gave way to caution, and the appetite for investing in unproven ideas diminished. What emerged was not just a different management style, but a deeper transformation in how teams operated and made decisions. Innovation became more constrained, filtered through layers of justification and risk analysis. Operational efficiency started to outweigh creative ambition. The dynamic, boundary pushing energy that had propelled Amazon's rise felt muted, replaced by a more corporate and restrained atmosphere that fundamentally altered how the company moved forward.

The change wasn't immediate, but it was palpable. Conversations that once revolved around bold "what if" scenarios became more focused on mitigating risk and protecting existing successes. Employees who had once been encouraged to challenge the status quo now faced greater scrutiny, and the margin for error seemed to shrink. The entrepreneurial spark that had attracted so many to Amazon dimmed, replaced by a growing sense of caution that ran counter to the company's original DNA.

This cultural shift marked the beginning of a more toxic undercurrent within Amazon. The relentless pursuit of efficiency and perfection came at the expense of the creative freedom and autonomy that had once empowered employees to thrive. What had been a company driven by fearless innovation began to feel like a machine, where risks were discouraged, and failure was met with consequences rather than lessons. The stark contrast between Bezos' bold leadership and Jassy's cautious pragmatism became increasingly apparent, reshaping the very fabric of Amazon's identity and leaving many employees grappling with a sense of disillusionment.

For a company that built its identity on thinking big, the turn toward conservatism felt like a quiet betrayal of its own origins. The transition under Andy Jassy was not simply a leadership change. It was a cultural pivot that altered how decisions were made, how risk was treated, and how people experienced the company day to day. For those of us who had been drawn in by the daring, boundary pushing ethos of the Bezos era, it was a sobering reminder of how quickly a culture can shift, and how those shifts do not always move in a healthier direction.

As I continued to advance, I found myself surrounded by more leaders who openly expressed dissatisfaction with what Amazon was becoming. Conversations that once focused on problem solving and innovation increasingly turned into venting sessions about burnout, pressure, and a growing sense of futility. Many admitted, sometimes quietly and sometimes bluntly, that they were actively looking for a way out. The frustration was not isolated. It was widespread, and it reflected something deeper than individual exhaustion. The culture had become harsher, less forgiving, and more punitive, even as the company continued to celebrate high standards and performance. Too often, those standards were enforced at the expense of the people expected to uphold them.

Nowhere was this more apparent than in the Weekly Business Reviews. WBRs quickly became the clearest expression of Amazon's cutthroat tendencies. These calls were dreaded by most, myself included. Leaders from stations across an entire region were pulled together to dissect the prior week's misses, bluntly labeled as failures. In theory, the meetings existed to drive improvement. In practice, they often became public dressing downs. Senior leaders and program managers would single out station managers and tear into them over performance, frequently without context or acknowledgment of constraints. There was little collaboration and even less coaching. Praise was rare. Constructive dialogue was rarer still. What remained was blame, delivered publicly, week after week, reinforcing a culture where fear replaced learning and survival mattered more than growth.

These sessions were intended to give leadership a clear snapshot of performance, but in practice they felt far more like interrogations than collaborative discussions. Leaders at every level dreaded them, knowing they would be placed under a microscope, with their decisions and results picked apart and questioned with little tolerance for context or nuance. The tone was sharp, the scrutiny unrelenting, and the pressure constant. It did not matter if you had delivered strong results for months. One bad week was enough to put your credibility at risk. The meetings were not structured to encourage growth or shared problem solving. They were exercises in assigning blame.

WBRs demanded an almost obsessive fixation on metrics, reinforcing a survival of the fittest mindset. Leaders were expected to defend every number, justify every miss, and produce corrective actions on demand, all while maintaining composure under aggressive questioning. The unspoken rule was unmistakable. If you could not withstand the pressure, you did not belong. While this environment may have pushed a small number of people to sharpen their performance, for most it bred fear and self preservation. Learning was secondary to damage control. Collaboration gave way to defensiveness. These meetings were not about improving the system. They were about proving you could endure the fire without breaking.

What struck me most about these WBRs was the complete absence of humanity in how people were treated. Metrics consistently mattered more than the individuals behind them, and the relentless fixation on numbers left no room for empathy or context. Leaders who raised legitimate concerns about associate burnout, staffing limitations, or operational constraints were routinely brushed off or told to simply "figure it out." The message was unmistakable. Results came first, and the cost to the people producing them was irrelevant.

Senior leadership's posture during these meetings only deepened that damage. Station managers were publicly criticized for performance misses, often without any meaningful guidance or practical solutions offered in return. The feedback was not constructive. It was punitive. Over time, this created a self reinforcing cycle of stress and emotional exhaustion. Even strong, capable leaders began to doubt themselves, not because they lacked skill or effort, but because they were operating in an environment that stripped away psychological safety. The WBRs did not just measure performance. They eroded confidence, morale, and the sense that leadership actually cared about the people carrying the weight of the operation.

For many leaders, WBRs became the breaking point. The constant pressure to defend every number in a combative setting took a real emotional toll, leaving people drained and increasingly disillusioned. Over time, those meetings started to feel like a condensed version of Amazon's larger culture, one that valued performance at any cost, often at the expense of the people delivering it. For those of us who remained, it grew harder to ignore the widening cracks. One by one, talented and capable leaders reached their limit and chose to walk away. In hindsight, it was clear that the environment created by WBRs was not just burning out individuals. It was quietly eating away at the very thing that made Amazon work in the first place, its people.

I still remember my time at the Rancho Sereno station, largely because of one especially toxic presence: Jack, the Fleet Program Manager. To this day, I couldn't tell you what qualified him for the role, although at Amazon that question often went unanswered. The company had a habit of elevating unqualified people into positions of influence, usually with predictable results. Even so, Jack managed to stand out. His behavior during Weekly Business Reviews was infamous, and his unprofessionalism became the defining feature of every meeting he touched. No one was ever safe from his outbursts, but one incident involving my friend Kassie, the station manager at Glenmoor, stands out as especially cruel.

During one WBR, Jack fixated on Kassie over a handful of issues that were both trivial and operationally irrelevant. He went after her over minor scratches on DSP vans and DMV registration tags that were nearing expiration. Anyone with even basic last-mile experience knew these were non-issues. Scratches on vans were unavoidable, no matter how careful drivers were. And the registration tags were handled by the leasing company, which had a habit of sending them at the last possible moment. Kassie had no control over either situation, but that didn't matter to Jack. He turned it into a personal attack, laying into her so aggressively that she broke down in tears in front of her peers and senior leadership.

The call went quiet, and no one stepped in. That was the power Jack held, not because he was respected, but because his behavior had gone unchecked for so long that it had become normalized. His tirades were not interruptions. They were expected, absorbed into the rhythm of those meetings as if they were just another agenda item. Kassie's humiliation was not an unfortunate lapse in professionalism. It was the system functioning exactly as it had been allowed to function. The moment left everyone shaken, but no one was truly surprised.

This wasn't an isolated incident. Jack's behavior was part of a broader pattern of bullying that ran straight through Amazon's work culture. The environment he and others like him created was toxic and demoralizing, grinding people down in plain sight. One of the few moments of resistance I ever saw came when a station leader, clearly at his breaking point, finally snapped and told Jack to fuck off during a call. For a split second, everything went silent. It felt like an unspoken acknowledgment that plenty of people were thinking the exact same thing. And then… nothing changed.

Jack's behavior continued exactly as before, completely unchecked. Upper leadership sat in on those calls, heard everything, and chose to do nothing. Their silence wasn't confusion or oversight. It was complicity. Eventually, Jack left

Amazon to start his own DSP, and honestly, I haven't given him a second thought since. Good riddance.

This culture of bullying wasn't confined to WBRs. It bled straight into daily operations, infecting every level of leadership. And if a leader decided they didn't like you, you were on borrowed time. They would try just about anything to undermine you or push you out. My manager at the Marwood Hills station, Scott, was one of the most capable and knowledgeable leaders I ever worked with, and even he wasn't immune to that toxicity.

On multiple occasions, station leadership pressured Scott to perform tasks he was not certified to do, including moving pallets out of a docked linehaul trailer. His TDR (Trailer Dock and Release) certification had expired, something leadership was fully aware of. Without a valid certification, stepping into that trailer would have meant immediate termination. Still, they kept pushing him to take the risk anyway. It was a blatant example of Amazon's willingness to ignore its own policies and put employee safety and job security second when it suited the operation.

Scott's situation captured the broader toxicity inside Amazon. Leaders were held to impossible standards while being actively undermined by the very people who were supposed to support them. The mix of bullying, negligence, and an obsessive focus on metrics created an environment where trust, collaboration, and basic humanity were pushed aside and replaced with fear and self-preservation. It was a system that didn't just wear people down. It broke talented leaders like Scott and exposed deep flaws in Amazon's work culture, flaws that have lingered and continue to stain the company's reputation to this day.

By the time I reached that point in my career, the pattern was undeniable. Amazon had not simply drifted away from its early culture. It had actively replaced it with something harder, colder, and far less forgiving. What once felt like principled intensity had curdled into intimidation. What had once rewarded curiosity now punished vulnerability. The same company that preached leadership principles about trust and ownership routinely tolerated behavior that eroded both. Watching leaders like Kassie humiliated, seeing Scott pushed toward impossible and unsafe choices, and hearing silence from those with the power to intervene stripped away any remaining illusions I had about how change really happened inside the system.

What made it worse was that none of this was accidental. These were not isolated bad actors slipping through the cracks. This was behavior reinforced through inaction, normalized through repetition, and protected by hierarchy. The message was clear. Performance mattered. People were expendable. If you could endure the pressure and stay quiet, you survived. If you pushed back, you became a problem to be managed out. Over time, that environment selected for exactly the kind of leadership that thrived on fear and control, while steadily driving out those who led with empathy, integrity, and perspective.

That realization marked a turning point for me. I had entered Amazon believing that high standards and humanity could coexist, that you could build something extraordinary without breaking the people doing the work. By this stage, it was clear how rare that balance had become. The system no longer bent toward growth or learning. It bent toward compliance. And once you see that clearly, once you understand that the culture itself is the constraint, you start asking different questions. Not how to succeed inside it, but how much of yourself you are willing to lose by staying.

When the Work Finally Felt Right

Coming back to the station where my Amazon journey first began felt both nostalgic and energizing. There was comfort in the familiarity, but it came with a renewed sense of purpose and the excitement of new challenges ahead. My new manager, Paul, quickly set the tone for what would become one of the most rewarding stretches of my time at Amazon. He was laid-back, in his mid-30s, and embodied what real leadership looked like. He trusted his team to do the work, avoided micromanaging, and kept communication open and honest. Whether we were talking through day-to-day operations or broader company goals, Paul's transparency made it easy to stay focused without constantly second-guessing the direction we were headed.

I was placed on the RTS front half and jumped straight into the operation alongside a group of sharp, driven professionals who were fully invested in tightening up Amazon's Last Mile process. Every day carried a real Day One energy, where problem-solving and innovation were not buzzwords but necessities. Many of the managers had worked their way up through the ranks the same way I had, and having crossed paths with some of them before, there was an immediate sense of trust and camaraderie. We shared the same scars from the grind: long, stressful hours, unrealistic expectations, WBRs, and exposure to incompetent, toxic leadership. It felt like being part of a team that understood the work, respected one another, and genuinely wanted to make things better.

The bond within our team was real, the kind that only forms when people have been through the same grind together. It reminded me of how Navy SEALs talk about the camaraderie that comes out of surviving BUD/S, not because the experience is glamorous, but because everyone knows exactly what the others endured. We trusted one another because we had all taken the same hits: long hours, constant pressure, and expectations that rarely matched reality. That shared understanding created a level of trust that made the team stronger, faster, and more effective. Everyone had each other's back, not out of obligation, but because we knew what it took just to get through the day.

That unity went beyond just running a clean operation. It made the work genuinely fulfilling. Every shift felt like a mission, driven by collaboration, purpose, and a collective push to do the job right. I thrived in that environment. The mix of teamwork, problem-solving, and a real Day One mindset kept me energized. There was something deeply satisfying about ending a shift knowing we had handled the chaos, hit the numbers, and still treated each other like humans. I was genuinely jazzed about the work and excited for each new shift. I actually looked forward to coming in, driven by the camaraderie and whatever challenges the next day was going to throw at us.

At that point, I was fully absorbed in Amazon's world, hooked by the culture, the pace, and how sharply it contrasted with my earlier experience at Marwood Hills. I talked about it constantly, to family, friends, even people who probably didn't care all that much. I shared stories about teamwork under pressure, the rush of solving operational problems, and the raw energy of the station. I'm sure my enthusiasm was a lot for anyone unfamiliar with Amazon's high-octane environment. But for me, it was never just a job. It felt like a community, something I took real pride in, and, maybe most dangerously of all, something I genuinely loved.

At the time, I genuinely believed everything was lining up. The struggles from earlier felt far away, replaced by confidence in Amazon's culture and where it was headed. The energy of the work was contagious. Even when my shift ran deep into the night, sometimes keeping me at the station until 2:30 a.m., I went to bed wired, already looking forward to getting back in. It wasn't just the work itself. It was the people, the systems, the constant problem-solving, and the camaraderie that made the brutal hours feel secondary. Being surrounded by others who brought the same intensity and commitment created its own momentum.

There's a rare kind of feeling that comes from pushing yourself harder than you ever have and still knowing you're exactly where you're supposed to be. That's what Amazon felt like in those moments. Every day brought new challenges, but also new chances to learn, tighten processes, collaborate, and contribute to something larger than any one person. There was an electricity in the air, a sense that we were part of something being built in real time. Every small win mattered. Every chaotic loadout and every problem solved reinforced that feeling of belonging and shared purpose.

Of course, it wasn't perfect. The pace was brutal and the expectations never let up. But in that moment, none of it seemed to matter. I was thriving in an environment that pushed me hard, rewarded ingenuity, and made me feel like I was part of something bigger than myself. Looking back now, I can see the cracks that would eventually widen into something unmanageable. At the time, though, all I felt was the rush of it. The long nights, the early mornings, the constant sense of purpose. It was intense, exhausting, and unforgettable. And for that stretch of time, I wouldn't have traded it for anything.

That stretch at Ashford felt like proof that Amazon could still get it right when the conditions lined up. Good leadership, shared experience, and mutual trust created a version of the job that felt sustainable, even meaningful. It showed me what was possible when pressure was paired with respect, when intensity did not have to come at the expense of dignity. Those nights left me tired but fulfilled, proud of the work and proud of the people beside me. It was the kind of environment that made you believe the system could evolve if enough of the right people were in the room.

But even then, beneath the energy and momentum, there was an unspoken fragility. This version of Amazon existed because of specific people, not because the culture demanded it. It depended on leaders like Paul, teams forged through shared hardship, and a narrow window where trust had not yet been eroded. I did not recognize it fully at the time, but that made it temporary by design. The system did not protect what worked. It tolerated it until it no longer did.

Looking back, that period stands out not just because it was good, but because it clarified the contrast. It showed me what I was fighting for without realizing it, and what I would later be asked to give up. In those moments, I felt aligned, energized, and certain I was exactly where I belonged. What I did not yet understand was how rare that alignment was inside Amazon, or how quickly it could disappear once the wrong pressures crept back in.

Work Hard, Play Hard, Deliver Harder

One of the more entertaining aspects of station life was something I jokingly called "Amazon Tinder." The high-pressure, fast-paced environment, paired with grueling hours that left little room for an actual personal life, naturally gave rise to workplace romances, especially in a station filled with a diverse group of people in their 20s. It wasn't surprising. When you spend nearly all your waking hours with the same people, bonds form quickly, and for some, those bonds turned personal. Relationships regularly blossomed between team members and even with DSP associates, creating a unique and sometimes amusing social dynamic inside the station.

The infamous Supply Room became the unofficial epicenter of those behind-the-scenes entanglements. Officially, it was nothing more than an operational space, a place to stash extra scanners, radios, and supplies. Unofficially, it carried a reputation that everyone knew about and no one ever talked about out loud. Its tucked-away location made it the perfect escape hatch from the floor, far enough removed from the constant motion and noise to feel private, even if it really wasn't. Ironically, I was one of the few managers who actually had access to the key code. Somehow, that code managed to spread anyway, passed along quietly, further cementing the room's legend.

What happened around that room was never just about hookups or flirtation. It was about proximity, exhaustion, adrenaline, and the strange intimacy that forms when people are thrown together under constant pressure. Long shifts blurred the lines between work and personal life. Late nights turned into shared decompression sessions, venting about leadership, laughing about disasters from the shift, or just stealing a few minutes of quiet away from the chaos. Some of those moments turned into relationships, some into short-lived flings, and some into messy situations that everyone pretended not to notice the next day. It was Amazon Tinder in its purest form, born out of necessity more than intention.

What always struck me was how seamlessly all of it coexisted with the work itself. No matter what drama simmered after hours or behind closed doors, once the floor lit up and the belts started moving, everyone snapped back into place. The station still ran like a machine. Metrics still mattered. Loadouts still had to clear. The unspoken rule was simple and universally understood: whatever happened off the clock stayed off the clock. Work hard, play hard, deliver even harder.

Looking back, that stretch at Ashford feels like a high point. It was a place where challenges still felt like opportunities, where camaraderie fueled innovation, and where the environment had a pulse that was both relentless and deeply human. It wasn't just a job. It was an ecosystem, one shaped as much by friendships, rivalries, crushes, and inside jokes as it was by KPIs and delivery metrics. The turbulence that would eventually change how I saw Amazon was already forming somewhere ahead, but in that moment, I was fully immersed in the rhythm of a workplace that felt alive, unpredictable, and driven by the flawed, complicated people inside it.

In hindsight, Amazon Tinder was more than just a running joke. It was another signal of how compressed life became inside the station. When work consumed nearly everything, it started absorbing the spaces where real life was supposed to exist. Relationships did not form because people were reckless. They formed because the job left little room for anything else. The station became the center of gravity, the place where stress, connection, release, and identity all collided. That intensity bonded people quickly, but it also blurred boundaries in ways that were easy to laugh about at the time and harder to unpack later.

What I did not see then was how fragile that ecosystem really was. The camaraderie, the humor, the shared exhaustion, all of it thrived only as long as the pressure stayed survivable and leadership stayed grounded. It felt alive because people were still allowed to be human inside it. But that balance was never protected by policy or culture. It existed only because of timing, personalities, and a brief alignment of incentives. Once those shifted, the same intensity that created connection would start grinding people down.

For that moment, though, it worked. We showed up hard, laughed harder, and carried the weight together. The station moved because people cared, not because they were afraid. That version of Amazon felt real, messy, imperfect, and alive. And that is exactly why it stayed with me. Because once you experience a workplace where pressure and humanity briefly coexist, you never stop noticing when one is sacrificed for the other.

The People Who Held the Middle

Tori was the back-half manager for RTS, and every Wednesday, what we called Cross-Over Day, our paths inevitably crossed. From the start, it was obvious he was someone solid. We clicked easily, bonding over things like parenting, travel, and food, probably helped by the fact that we were close in age and at similar points in life. Beyond that personal connection, Tori brought real depth from his time at Amazon. He understood the operation inside and out, and that experience consistently showed up in smarter processes and smoother execution.

What really set Tori apart, though, was how he led. He managed to be firm without being rigid and supportive without being soft. He created space for open communication and genuinely encouraged people to speak up, which made his team feel heard instead of managed. He knew how to match work to people's strengths, putting associates in positions where they could actually succeed, and when guidance was needed, he gave it without ego or theatrics. That balance-built trust quickly and kept the team aligned.

Tori was also intentional about recognizing good work and addressing issues head-on without turning them into personal attacks. Wins were acknowledged, problems were handled clearly and fairly, and no one was left guessing where they stood. His approach to problem-solving blended decisiveness with collaboration, pulling people in rather than shutting them down, even when the issues were complex or time-sensitive.

That mix of adaptability, empathy, and competence made Tori not just a strong leader, but someone people genuinely wanted to work with. He built an environment where the team felt energized instead of drained, and where difficult work felt manageable instead of overwhelming. Working alongside him was one of the highlights of my time at Ashford, and his leadership left a lasting mark on how I think about teamwork, trust, and what good management actually looks like in practice.

Among my direct reports, one person stood out immediately: Alejandro. He was young, high-energy, and relentlessly curious. He didn't have much in the way of formal education, but what he lacked on paper he more than made up for with drive and an almost obsessive hunger to learn. Mentoring Alejandro quickly became one of the most rewarding parts of my role. He soaked up everything, from core operational processes to advanced Excel skills like Pivot Tables, and he didn't stop at understanding how things worked. He applied what he learned. Before long, he was building analytical reports on his own, flagging potential system failures before they had a chance to blow up. Knowing he was on the floor gave me a rare sense of calm, because I trusted that he had things handled.

What really separated Alejandro from the rest, though, was his instinct for leadership. It wasn't forced or performative. It just showed up naturally. Whether he was managing associates or coordinating a Crash Sort, he had an intuitive sense for organizing people and work. More than once, I walked into a situation ready to outline a plan, only to find that Alejandro had already assigned roles and gotten the operation moving. His initiative and ability to earn buy-in from his peers made him indispensable, not just to the team, but to the operation as a whole.

Seeing that potential, I became fully invested in his growth. I pushed hard for his promotion and made sure he was ready when the opportunity came. I coached him through the interview process, helped him sharpen his responses, and showed him how to clearly articulate his impact instead of downplaying it. When the interview finally happened, he crushed it. He was offered a management role, along with the perks that came with it, and it felt like one of those rare moments at Amazon where the system actually worked the way it was supposed to.

To my surprise, Alejandro turned the offer down. When he explained why, it caught me off guard. He told me it wasn't about fear of responsibility or lack of ambition. It was about trust. He said the mentorship we'd built mattered to him, that working under my guidance had shaped his growth more than any title ever could. He also knew I was likely moving on and was honest about his concern that, in a new role, he could easily end up reporting to a toxic leader and walking straight into the kind of stress he had watched burn people out. Hearing that hit hard. It was one of those moments that quietly confirms the kind of impact you hope you're having, even when the system rarely rewards it.

I encouraged him to think about the long-term opportunities Amazon could offer, but his mind was already made up. Amazon's structure doesn't really allow associates to follow managers into new roles, and eventually our paths split. When I moved on, Alejandro made the decision to leave Amazon as well.

I still think about him from time to time and wonder where life took him next. I hope he found a place that recognized his ability, protected his curiosity, and gave him room to grow without grinding him down. Alejandro's story has stayed with me because it captures what mentorship is really about. It's not just teaching skills or preparing someone for the next role. It's building trust, confidence, and a sense of self-worth that carries forward long after the job ends. In a system that so often treats people as interchangeable, that kind of connection feels rare. And it's a reminder that investing in people can leave a legacy that outlasts any title or promotion.

Working with people like Tori and Alejandro reinforced something I had already begun to understand, but had not fully named yet. The best parts of Amazon were never the systems, the scale, or the slogans. They were the individuals who showed up with competence and empathy at the same time, who made the work feel human without sacrificing standards.

When leadership worked, it worked because people trusted one another, not because a process demanded compliance.

Those relationships also exposed the fragility of that success. Tori's leadership and Alejandro's potential thrived only within a narrow window where the environment allowed it. Neither of them needed more pressure or more oversight. They needed protection from the wrong kind of leadership, something Amazon rarely provided. Watching someone like Alejandro walk away rather than risk falling under a toxic manager was not a failure of ambition. It was a rational response to a system that had taught him exactly what it rewarded and what it destroyed.

By then, the pattern was unmistakable. The people who made Amazon better were often the ones quietly planning their exit, not because they lacked resilience, but because they had enough clarity to see what staying would eventually cost them. Mentorship, trust, and shared growth created real impact, but they existed in spite of the culture, not because of it. And as more people like Tori and Alejandro passed through my life, it became harder to ignore the truth. Amazon did not struggle to find talent. It struggled to keep the kind that actually mattered.

Your Profit Matters

As my time at the Ashford station progressed, I became deeply involved in onboarding new Delivery Service Providers, many of whom were first-time entrepreneurs stepping into Amazon's unforgiving Last Mile ecosystem. For most of them, this wasn't just a new contract. It was a financial and personal gamble. They were investing their savings, taking on debt, hiring employees, and trusting that Amazon's model would actually work for them. I understood that risk intimately, having lived it myself as an entrepreneur, and I took their success personally.

From day one, I was clear about Amazon's performance expectations. I didn't sugarcoat the metrics or the pressure that came with them. But I also refused to treat DSP owners like disposable operators in a rigid system. I spent time walking them through the reality of route efficiency, staffing models, safety compliance, and on-time delivery, not as abstract KPIs, but as levers that directly impacted their profitability. We talked openly about costs, margin erosion, turnover, and the brutal math of running routes day after day. My focus wasn't just on whether they could survive Amazon's standards, but whether they could actually build something sustainable within them.

Early in those conversations, I was deliberately blunt. I would tell them, "The only thing that matters to me is that you make money." That statement often landed with skepticism. Most of them were conditioned to hear corporate language about customer obsession and metrics, not someone inside Amazon openly prioritizing their bottom line. But over time, they saw I meant it. I wasn't interested in pushing compliance for compliance's sake. I was invested in helping them understand where Amazon's systems helped them and where they quietly worked against them, so they could make smarter decisions and protect their business.

I pushed them to think beyond the day-to-day chaos of dispatch and rescues. We talked about retention strategies, training investments, route planning discipline, and knowing when to push back instead of blindly absorbing every operational change. I encouraged them to look at their business over quarters and years, not just weekly scorecards, because I knew how quickly burnout and financial strain could wipe out even the most motivated owner. In a program that often-treated DSPs as interchangeable, I tried to give them the clarity and confidence to operate like real business owners.

Working alongside these DSPs reignited something in me. Watching them move from overwhelmed and uncertain to confident and capable was deeply satisfying. When they hit milestones, stabilized their operation, or finally saw consistent profitability, it felt like a shared win. Their progress validated the time and energy I put into them and reinforced my belief that mentorship matters, especially in environments that rarely prioritize it.

That chapter of my career also pulled me back to my own entrepreneurial roots. I saw my past self in many of those owners: the risk-taking, the sleepless nights, the constant problem-solving, and the quiet pride that comes from building something from the ground up. Helping them succeed wasn't really about packages or routes. It was about empowering people to navigate a brutal system with their eyes open and giving them a fighting chance to come out ahead. That, more than any metric, made the work meaningful.

Over time, that role sharpened a hard truth about Amazon's DSP model. The company spoke about partnership, but the structure was built around asymmetry. DSPs carried the financial risk, the labor exposure, and the day-to-day volatility, while Amazon retained the leverage and the ability to change the rules without absorbing the fallout. Success inside that model required more than hustle or compliance. It required clear-eyed realism, strong leadership, and someone willing to translate corporate expectations into survivable business decisions. Too often, that translation never happened.

Being able to play that role gave me purpose, but it also deepened my unease. I could help individual owners navigate the system, but I could not change the system itself. For every DSP that stabilized and found its footing, there were others quietly bleeding cash, burning out, or one bad quarter away from collapse. Amazon measured DSP health through dashboards and scorecards, but the real story lived in late-night phone calls, stressed conversations about payroll, and owners wondering how much longer they could hold on.

That tension stayed with me. The work was meaningful because it was human, but it was also a reminder of the limits of influence inside a machine designed to protect itself first. Mentorship could soften the edges, but it could not fix the imbalance. As I moved deeper into that space, it became harder to ignore that helping people survive inside Amazon was not the same as believing in the system they were trapped in. And that realization would eventually force me to confront where my own line was, and how long I was willing to stand on the inside translating risk for everyone else.

Amazon's Strategic Misstep

During this period, I watched Amazon make what I still believe was one of its most damaging strategic decisions, especially for the DSP program. Roughly six months ahead of Peak, Amazon rolled out a new delivery model that introduced Amazon-employed drivers into last-mile delivery and was framed as a short-term pressure valve, a temporary solution to help absorb the holiday surge. That explanation never fully passed the smell test. It became clear very quickly that if the model worked operationally, it was not going anywhere.

What Amazon either failed to acknowledge or chose to ignore was how directly this move undercut its DSP partners. DSP owners had been encouraged to scale up aggressively. They invested heavily in vans, hired and trained drivers, expanded fleets, and structured their entire businesses around the assumption that Peak volume was how they made the math work. The holiday season was not just busy. It was the period that carried them through slower months and kept their businesses solvent. The new delivery model quietly siphoned that volume away.

Routes that DSPs were counting on were suddenly reassigned to the new drivers. That meant fewer deliveries, fewer hours, and less revenue, all while fixed costs stayed exactly the same. Vans still had to be paid for. Insurance still came due. Payroll did not magically shrink just because Amazon decided to experiment. For many DSPs, the margins were already razor thin. This new delivery model pushed some of them right up against the edge.

The frustration from DSP owners was constant and raw. I heard it every day. These were not abstract complaints. These were people staring at spreadsheets, realizing the numbers no longer worked, and feeling trapped. They had built their businesses around Amazon's model because Amazon told them to. This new delivery model felt like the rug being pulled out from under them without warning. For owners who had poured their savings, time, and identity into these operations, it felt like a betrayal, plain and simple.

What made it worse was the lack of transparency. Amazon continued to talk about partnership while quietly reallocating routes. DSPs were expected to absorb the hit without complaint, adjust on the fly, and somehow remain profitable. There was no meaningful runway, no real compensation, and no acknowledgment of the damage being done. The risk was entirely pushed downstream, exactly where Amazon claimed it was not.

From my position, there was very little I could do to change the outcome. I listened. I tried to help owners rethink staffing, routes, and cost controls. I offered whatever guidance I could within the limits of my role. But no amount of coaching could fix a model where volume was being deliberately diverted away from the very partners Amazon depended on. The damage was structural.

The new delivery model did not strengthen Amazon's last-mile network. It fractured it. More importantly, it broke trust. DSPs who had believed they were building something sustainable suddenly realized how expendable they really were. This new delivery model exposed a hard truth. When it came time to choose between protecting its partners or protecting its own incremental profitability, Amazon chose itself without hesitation. That decision left lasting scars across the DSP community, and many never fully recovered from it.

The model itself was riddled with problems from the moment it launched. Amazon rushed to hire, train, and deploy its own fleet of drivers, complete with branded vans and dedicated routes, selling the roles as seasonal positions with the promise of possible full-time employment after Peak. On paper, it sounded clean and controlled. In practice, it was anything but. The rollout was hurried, the vetting process was shallow, and training was bare minimum at best.

Many of these drivers were young, some fresh out of high school, others with little to no prior work experience, let alone experience in a high-pressure logistics environment. They showed up eager, but eagerness doesn't compensate for preparation. Last Mile delivery is unforgiving even for seasoned drivers. Throwing undertrained employees into Peak season routes was a recipe for failure, and it showed almost immediately.

Driver performance became a daily battle. It was not uncommon to see drivers returning to the station hours early after delivering only a fraction of their route. In cases where drivers were willing to listen, I sent them back out to finish the work they had signed up to do. Others flat-out refused. Instead, they dumped their undelivered packages into what we called a crash sort, essentially pushing their workload onto Amazon Flex drivers to clean up.

That behavior quickly spiraled out of control. Crash sort volume exploded from a manageable 700 packages to an unworkable 6,000 to 7,000 packages in a single day. Even when we tried to compensate by bringing in more Flex drivers, there simply were not enough available to absorb that kind of overflow. Packages piled up, routes went uncovered, and Promised Delivery Dates started slipping. What was supposed to be a support mechanism turned into a bottleneck that threatened the entire station's performance.

Safety issues made everything worse. Accidents involving these drivers became routine. Vans were T-boned at intersections. Roofs were peeled back after drivers misjudged parking structures. Parked cars in customer driveways were hit. Curbs, poles, fences, nothing was safe. Nearly every day, I was logging new incidents into Amazon's Workplace Health and Safety system, Gensuite. Accident reporting stopped feeling like an exception and started feeling like part of the daily checklist.

The combination of poor training, inexperience, and unrealistic expectations turned this into both an operational and safety nightmare. Instead of easing Peak pressure, the delivery model created more work, more risk, and more damage, all while undercutting the DSPs who had spent years learning how to do the job the hard way. Watching it unfold in real time, it was impossible to ignore the irony. Amazon had built an entire partner ecosystem to handle Last Mile delivery, then sidelined it for a rushed experiment that failed the people, the operation, and ultimately the customer.

Facing chaos, I turned to the DSPs for help. Leveraging my strong relationships with the owners, I proposed scheduling additional routes later in the day to address the shortfalls. Despite their initial frustration with the new delivery model, the DSPs saw the financial upside of taking on more routes and agreed. Securing leadership buy-in for this shift in RTS (Return-to-Station) operations was another challenge. I proposed conducting a mid-day crash sort to redistribute undelivered packages and creating additional DSP routes in the afternoon. While it required meticulous planning and considerable effort, this approach ultimately proved successful.

The new plan worked seamlessly. DSPs earned more money, customers received their packages on time, and leadership hit their performance metrics. What could have been a logistical disaster during Peak was turned into a functional, if not flawless, operation. However, in true Amazon fashion, the solution went unacknowledged. There were no thanks or recognition for the effort it took to resolve the crisis.

Still, the experience underscored the resilience and adaptability of the DSPs and the teams on the ground. While this new delivery model had threatened to derail the season, the collaboration and determination of those involved ensured we turned chaos into success. Yet, it also left a lingering question: how many more strategic missteps could Amazon afford to make before undermining the trust and loyalty of its most vital partners?

When Peak finally passed, the numbers told a familiar Amazon story. Volume moved. Metrics recovered. Leadership declared success and moved on. But the people who carried the weight of this new delivery model did not forget what it cost them. DSP owners remembered the routes they lost, the margins that vanished, and the silence that followed. Associates remembered the chaos, the safety incidents, and the extra work dumped on them to compensate for a model that never should have launched the way it did. The system absorbed the damage and kept moving, but the trust that was lost never came back the same.

For me, this stripped away any remaining ambiguity about how Amazon viewed partnership. When conditions were favorable, DSPs were framed as essential collaborators. When conditions tightened, they became optional buffers, easily sidelined to protect corporate flexibility. The language never changed, but the behavior did. Watching that disconnect play out in real time made it harder to keep translating Amazon's decisions into something DSPs could live with. There were limits to how much risk could be pushed downstream before the relationship broke.

That period marked another quiet shift in how I saw my role. I could still solve problems. I could still stabilize chaos. But I was increasingly doing it inside a framework that created the very fires it expected us to put out. This was not just a flawed delivery model. It was a signal. A reminder that operational success did not equal ethical alignment, and that loyalty inside Amazon flowed in one direction only. Once that reality became clear, it was impossible to unknow it.

Saving Routes in Real Time

One of the parts of working at Amazon that I genuinely respected was the sophistication of its on-road technology. When it worked as intended, it gave unparalleled visibility into Last Mile operations and allowed problems to be identified and addressed before they spiraled out of control. Amazon's systems were built to surface issues in real time, not hours later when the damage was already done. That capability fundamentally changed how delivery operations could be managed.

The most impactful system introduced during my time was a real-time delivery tracking platform that replaced the earlier workarounds we had used to monitor DSP performance. Before that, I had developed my own method to track delivery progress because the existing visibility was limited. The new platform was a significant step forward. It provided a live, end-to-end view of DSP performance across the station, breaking down delivery progress at both the company and individual driver level. At a glance, I could see which DSPs were moving smoothly, which were struggling, and which individual Delivery Associates were on pace to miss their routes entirely.

The platform made the invisible visible. We could tell if a driver was falling behind early in the day, stuck in a difficult delivery zone, or simply not moving at the expected pace. That level of insight allowed us to intervene proactively instead of reacting at the end of the shift, when options were limited and customer promises were already at risk. While DSP owners and managers technically had access to the same data, many were hesitant to fully embrace it at first. Some did not trust the system. Others did not want to confront what the data revealed.

I made it a point to actively monitor driver performance throughout the day and reach out to DSP owners or managers when I saw trouble brewing. If a driver was clearly falling behind, I would recommend initiating a Rescue. A Rescue meant sending a Delivery Associate who had finished, or was close to finishing, their own route to take a portion of undelivered packages from a struggling driver. Done right, it was one of the most effective tools for protecting both customer experience and DSP performance.

Rescues were not popular initially. Many DSP owners saw them as wasted labor. From their perspective, assigning drivers without full routes felt inefficient and expensive. I pushed back hard on that thinking. I explained that every package delivered still generated revenue and counted toward their DSP Scorecard. Leaving packages undelivered was far more costly than paying a Rescue driver for a few extra hours. Once owners started connecting the dots, especially after seeing Scorecard impacts, attitudes began to shift.

Over time, smarter DSPs began building Rescue drivers into their daily staffing plans. Assigning one or two DAs as dedicated Rescues gave them flexibility and insurance. When delays happened, and they always did, they could respond quickly instead of letting small problems snowball into missed deliveries, customer complaints, and metric hits. That adaptability often made the difference between a clean day and a disaster.

The DSP Scorecard was the gravity behind all of this. It was the single most powerful lever Amazon used to shape DSP behavior. Weekly ratings ranged from Fantastic Plus down to Poor, and those labels carried real financial consequences. A Fantastic Plus rating came with bonuses and first rights to additional routes, which directly translated into higher revenue. Anything below that meant fewer opportunities, reduced bonuses, and increased scrutiny. A string of Poor ratings triggered formal performance reviews, and I watched more than a few DSPs lose their contracts entirely because they could not recover.

The real-time tracking system and the performance scorecard worked together as both a guide and a threat. They enforced accountability, but they also gave owners a clear roadmap for improvement if they were willing to engage with the data honestly. Watching DSPs evolve was one of the more rewarding parts of the job. Owners who initially resisted the data eventually learned how to use it, tightening their operations, stabilizing routes, and steadily improving their performance standings. When that happened, it did more than improve metrics. It improved morale, predictability, and the long-term viability of their businesses.

Everything I did in this space was driven by Amazon's core Leadership Principle of Customer Obsession. For me, that principle wasn't abstract. It meant intervening early, pushing uncomfortable conversations, and holding drivers and DSPs to high standards because the customer on the other end of the delivery deserved it. The results spoke for themselves. Across the DSPs and the new drivers I directly managed, delivery completion consistently hovered around 99.9%. That wasn't luck. It was the product of using the technology aggressively, making hard calls in real time, and never losing sight of the customer experience, even when it made the operation uncomfortable for everyone involved.

What became clear over time was that technology at Amazon was never neutral. Real-time tracking systems had the power to elevate performance or intensify pressure, depending entirely on how leadership chose to use them. In the right hands, visibility created opportunity. It allowed for early intervention, smarter decisions, and outcomes that protected both customers and DSPs. In the wrong hands, the same data became a weapon, another way to monitor, threaten, and punish without context. The system itself was impressive. How it was applied was the variable that mattered.

That distinction shaped how I approached the work. I never saw real-time delivery data as a way to catch people failing. I saw it as a way to keep failure from happening in the first place. The difference may sound subtle, but on the floor, it was everything. When data was used to support rather than shame, DSPs leaned in. When it was used to intimidate, they shut down or cut corners. The technology did not create accountability on its own. Leadership did.

Looking back, this was another example of Amazon's core tension. The company built extraordinary systems but often lacked the restraint and empathy to deploy them responsibly. Customer obsession was real, but it was too often pursued through pressure instead of partnership. For a while, I believed that if the tools were strong enough and the intent was right, the culture could bend in a better direction. Over time, I learned that tools do not change culture. People do. And when the wrong incentives dominate, even the best technology can become part of the problem rather than the solution.

Finally in the Driver's Seat

Paul stayed true to his word, and before long I found myself in another POD. This time, I passed without issue. The rush of finally earning the promotion was real. After months of frustration, I had broken through and moved to the next level. That said, the reality behind the title set in quickly. The pay bump was minimal, the additional stock options were nice but far from life-changing, and my responsibilities exploded overnight. My days routinely ran 13 to 16 hours, and on the worst stretches, I pushed close to 20 hours straight. Strangely enough, none of it rattled me. I thrived in the chaos.

It was around this time that I started noticing something else. I was being paid significantly less than newer hires and younger managers doing the exact same job. I couldn't help but wonder whether age played a role, whether there was some unspoken message behind the numbers. The thought lingered, uncomfortable but hard to pin down. In the end, I chose to set it aside and keep moving forward, telling myself the work would speak for itself.

What made the workload bearable, what made it worth it, was the environment. I was surrounded by a strong leadership team, some of the best people I had worked with at Amazon, and no two days ever felt the same. Every problem was an opportunity to think, adjust, and improve something in real time. One small but surprisingly meaningful perk of the role was being issued an Amazon credit card. I used it not just for station expenses, but intentionally for morale. Snacks for associates, the occasional team lunch, small gestures that showed people they were seen. It was never listed in any job description, but I understood that in an environment that demanded everything from people, those little investments mattered more than leadership liked to admit.

For the first time, I also had the platform to lead in ways that actually felt meaningful. I created promo documents to help associates move into new roles, spent hours deep in the data, and leaned hard into using technology to drive operational improvements. The metrics followed. My numbers finally spoke loudly enough to get leadership's attention, and the recognition I had been chasing for years started to materialize. But my impact went beyond scorecards. I wrote and implemented multiple Standard Operating Procedures, many of which are still in use today, and authored whitepapers designed to tighten processes and give teams clearer paths to success.

One of my most important contributions was a whitepaper focused on reorganizing the Driver Training Program. I knew exactly where the gaps were and what needed to change to make the program more effective. The hard part was never the solution. It was convincing leadership to listen. At Amazon, good ideas were not enough. You had to sell them, defend them, and back them up with data that could survive scrutiny. That challenge did not intimidate me. It energized me.

To their credit, station leadership recognized just how much managers were being asked to carry. That acknowledgment sometimes showed up in practical ways, like allowing team lunches or dinners to be expensed without excessive pushback. It was a small gesture, but it reinforced the idea that effort was being noticed, even if the workload itself never slowed down. I thrived in that environment, fueled by the sense that my work mattered and that I was finally operating at the level I had been pushing toward.

And then there was the Amazon swag. Peak seasons and milestone events came with a steady stream of t-shirts, hoodies, cups, pins, and other branded gear. It might sound trivial, but it added to the energy. It created a sense of pride and shared identity, something tangible that made the grind feel a little lighter. I imagined managers across other stations experiencing the same thing, all plugged into the same rhythm, contributing to a culture that, at least in those moments, felt appreciative, energized, and alive.

That period felt like standing in the center of the machine with the lights finally on. I had the title, the access, and just enough influence to see how things really moved. The long hours did not bother me because the work still felt purposeful, and the trust from leadership made the grind feel earned rather than imposed. For a moment, it seemed like the system was finally rewarding effort instead of merely extracting it. I believed I had reached a place where competence, creativity, and endurance could coexist without constant friction.

But even in that momentum, there were early signals I chose not to dwell on. The imbalance between responsibility and compensation, the quiet normalization of sixteen-hour days, the idea that thriving meant being perpetually available. Those realities were easy to rationalize when the work was stimulating and the team was strong. Success inside Amazon has a way of masking its own cost, especially when you are wired to push through and deliver.

Looking back now, that stretch stands out as both a high point and a warning. It showed me what I was capable of when given room to operate, and it showed me how quickly overextension could be reframed as commitment. The system had not changed. I had simply learned how to run faster inside it. And while it felt sustainable in the moment, that pace would eventually demand a reckoning, one I was not yet ready to confront.

System Fails People

In February, just as the dust from Peak was starting to settle, upper leadership made a decision that sent shockwaves through the station. All the new Delivery Associates were terminated. No warning. No transition plans. Just a sudden, sweeping cut that left dozens of people without jobs overnight. The mood inside the building shifted instantly. What had been a place catching its breath after Peak turned heavy and tense, filled with confusion, anger, and disbelief. These were associates who had just pushed through the most demanding stretch of the year, and now they were being cut loose as if their effort had an expiration date. The way it was handled felt cold and impersonal, as though their contributions had been nothing more than a temporary convenience.

I was put in an impossible position. On one side, I was a confidant, listening to the frustration, fear, and heartbreak from people I had worked closely with and come to respect. These were solid performers who had done exactly what was asked of them and more. On the other side, I was still a manager, expected to keep the operation moving, maintain morale, and absorb the fallout without openly challenging the decision. All of it happened under the watchful eye of leadership, who seemed more focused on execution than the human damage left behind. My own anger sat just below the surface, but I knew dwelling on the injustice wouldn't help anyone. If I couldn't change the decision, I was going to do whatever I could to soften the landing.

I immediately leaned into the relationships I had built with DSP owners. I picked up the phone, sent messages, and advocated hard for the best drivers I knew. I spoke plainly about their work ethic, reliability, and performance, making it clear that these were people worth hiring. At the same time, I made myself available to the displaced DAs, setting up informal conversations, walking them through their options, and pointing them toward DSPs that were actively hiring. I helped connect the dots, offering guidance where I could and reassurance where I couldn't.

Watching those conversations turn into job offers was one of the few bright spots in an otherwise ugly situation. Seeing the relief on their faces when they realized they weren't starting from zero mattered. In a system that had discarded them without hesitation, I was able to help restore some sense of dignity and momentum. It didn't erase what had happened, but it reminded me that leadership still meant showing up for people, especially when the company they worked for chose not to.

Yet even with those small wins, the experience was bittersweet. Amazon made a half-hearted attempt to soften the fallout by creating a handful of new roles at the station for some former delivery associates, but it was obvious from the start that most people were going to be left behind. I couldn't shake the frustration of knowing that some were walking out with no options, no safety net, and no clear path forward. As hard as I pushed and as much as I tried to help, there were limits to what I could do. That sense of powerlessness lingered long after the initial shock wore off.

The whole ordeal left me carrying a complicated mix of pride and disappointment. I was proud of how I showed up, of the trust I had built, the doors I had helped open, and the tangible difference I made for the people I could reach. But watching others, people who had given everything they had during Peak, leave empty-handed was hard to stomach. It was a blunt reminder of how unforgiving the corporate world can be, where job security is fragile and loyalty rarely buys you protection.

More than anything, it crystallized a truth I had been circling for a long time. When systems fail people, all that remains is how we treat one another in the fallout. And in that moment, it became painfully clear that Amazon did not care about the human element of the operation. What mattered was efficiency, flexibility, and profit. People were interchangeable.

That realization stuck with me, and it would continue to shape how I saw the company long after the dust finally settled.

That decision stripped away any remaining ambiguity about where the line was drawn. Amazon could talk endlessly about culture, safety, and people-first values, but when convenience demanded it, those words evaporated. The delivery associates were not phased out thoughtfully or transitioned with care. They were erased. The speed and finality of it sent a message far louder than any leadership principle ever could. Contribution was temporary. Commitment was conditional. And once usefulness expired, so did concern.

What stayed with me most was not the anger, but the clarity. I had spent years believing that if you worked hard enough, cared enough, and led the right way, the system would eventually meet you halfway. Watching those terminations unfold made it impossible to maintain that illusion. The machine did not bend for effort or sacrifice. It adjusted only for efficiency. The humanity that showed up in the aftermath did not come from Amazon. It came from individuals filling a void the company chose to ignore.

That moment marked a quiet but decisive shift in me. I was still doing the work, still solving problems, still protecting the operation. But something fundamental had changed. I no longer believed that Amazon's success and human decency were aligned goals. From that point on, every decision, every long night, every request to absorb more would be filtered through a new question. Not whether it could be done, but whether it should. And once that question enters the room, it never really leaves.

I Trained Them, the System Weaponized Them

By this point in my Amazon journey, leadership had come to see me as the domain expert in Last Mile operations. That reputation made me the default trainer for every new manager and leader who came into the station. While it was validating to be trusted at that level, it also added another heavy responsibility to an already overloaded role. I took it seriously. I made a point to give each new leader the most thorough training I could, walking them through the tools, processes, and expectations that actually mattered on the floor, not just what looked good on paper.

To make the training more effective, I built my own PowerPoint decks specifically for onboarding new leaders. The goal was clarity, breaking down complex systems in a way that made sense in real operational terms. Long after the formal training ended, many of those managers continued to reach out to me for help, asking questions about processes they were still trying to understand or situations they hadn't encountered before. I never turned them away. Supporting them wasn't just part of the job at that point, it was part of how I operated.

Among the leaders I trained was Mark, a newly hired MBA graduate who arrived at the station with what appeared to be the right attitude and a genuine eagerness to learn. He was sharp, picked things up quickly, and on the surface seemed invested in understanding how the operation actually worked. I put significant time into training him, walking him through the nuances of our systems, night sort operations, Dispatch, RTS and the realities that don't show up in onboarding documents. We built what I believed was a solid working relationship, and I trusted him with insights that came from years on the floor, not from a textbook.

That trust turned out to be misplaced. Over time, Mark changed. The curiosity gave way to arrogance, and collaboration was replaced by intimidation. He became a bully, plain and simple, and it didn't take long to realize he was actively working to undermine my position. Watching someone I had invested so heavily in pivot toward self-preservation and power plays was hard to stomach. It wasn't just professional disappointment. It felt personal. I had helped him find his footing, only to later see him use what he'd learned as leverage against the very people who had supported him.

Grace joined around the same time. Like Mark, she was an MBA hire, and like Mark, she required extensive training to understand how Amazon actually operated beyond theory and dashboards. I spent countless hours with her, answering questions, explaining workflows, and helping her navigate the realities of the job. At the time, I believed we had built a collaborative, respectful working relationship.

A few years later, that illusion shattered during a Weekly Business Review. Grace, now promoted, launched into a vicious critique of a manager I worked closely with. It wasn't constructive. It wasn't professional. It was a public teardown. The intensity of her attack left the manager visibly shaken, and the silence that followed spoke volumes. No one stepped in. No one slowed it down. The meeting moved on as if nothing had happened.

The impact was immediate and lasting. That manager left the station shortly afterward, emotionally wrecked and burned out by an environment that allowed that kind of behavior to pass as leadership. Sitting there, watching it unfold, I couldn't help but reflect on how often this pattern repeated itself at Amazon. People were trained, promoted, and then rewarded not for lifting others up, but for tearing them down in rooms where power mattered more than decency.

Those moments stayed with me. They forced me to confront an uncomfortable truth about the culture I was helping sustain. Technical skill and intelligence were never the real problem. Character was.

That incident was far from isolated. It was part of a broader and deeply troubling pattern within Amazon's leadership culture. More and more, I heard stories of managers pushed to their breaking point by relentless pressure, public humiliation, and an almost complete absence of support. Some walked away on their own, choosing sanity over survival in a system that demanded everything and gave very little back. Others did not get that chance. I learned of one manager who took his own life, overwhelmed by the emotional strain and unrelenting expectations placed on him. That news landed hard. It stripped away any remaining illusions I had about how dangerous this environment could become when humanity was treated as optional.

I had always tried to build a space rooted in trust, support, and collaboration. I believed leadership meant protecting people, not grinding them down. But it became impossible to ignore how many were struggling quietly, carrying stress and fear alone because speaking up often came with consequences. The culture rewarded results, not restraint. It tolerated abuse as long as metrics were met. Toxic behavior was allowed to flourish, especially when it came from people with titles or influence.

That realization left me deeply conflicted. I was proud of the people I had helped, the careers I had supported, and the teams I had built. At the same time, I was increasingly disturbed by the system I was operating within, one that allowed so much damage to occur behind the curtain of performance and success. To me, it was heartbreaking that this was happening inside a company as admired and influential as Amazon.

What I did not realize then was that I was about to be pulled into that same gauntlet myself. The pressure, the scrutiny, the psychological toll. It would soon become personal, and it would test my limits in ways I never could have anticipated.

Looking back, that period forced a reckoning I had been postponing for far too long. I had spent years building leaders, transferring knowledge, and trying to model a version of leadership rooted in competence and care. Yet I was watching, in real time, how the system rewarded an entirely different behavior set. The people who absorbed knowledge and used it to protect teams rarely advanced. The ones who weaponized it, who learned how to perform dominance in the right rooms, climbed quickly. That wasn't a flaw in execution. It was a feature of the culture.

What disturbed me most was not betrayal by individuals like Mark or Grace. It was how predictable their evolution was once incentives shifted. Amazon didn't corrupt people randomly. It shaped them. It took smart, capable leaders and quietly trained them to value optics over outcomes, power over protection, and survival over responsibility. The cruelty wasn't always loud. Sometimes it showed up as silence, as a room full of leaders choosing not to intervene, choosing not to see. That silence did as much damage as any outburst.

By then, I was carrying more than operational weight. I was carrying the emotional residue of watching good people disappear, either burned out, pushed out, or broken down. I could still teach processes. I could still stabilize chaos. But I was beginning to understand that the deeper problem wasn't a lack of training or talent. It was a culture that treated psychological harm as acceptable collateral as long as the numbers stayed green.

When I said I didn't realize the gauntlet was about to close in on me, that was true in the narrow sense. I couldn't see the specific shape it would take. But the warning signs were everywhere. I had watched the pattern repeat often enough to know how it ended. The only real question left was whether I would recognize the moment when it was my turn, and whether I would have the clarity and strength to respond before the system took more than I was willing to give.

Stretch Roles and Broken Margins

Whenever a station in the region was struggling or short on leadership support, I was often tapped to step into what Amazon called a "stretch role." In practice, that meant juggling my responsibilities at my home station while parachuting into another operation that needed immediate stabilization. Over time, I supported every Distribution Center in the region, not because it was convenient, but because leadership knew I could walk into chaos and make sense of it. These assignments were exhausting, but they gave me a front-row seat to how wildly different stations could be, and how uneven the burden placed on DSPs often was.

One station I was repeatedly sent to sat in the heart of South-Central Los Angeles. The area had its own pulse, its own energy, and its own set of realities that no dashboard or leadership deck could ever fully capture. The station served dense neighborhoods, high-risk delivery zones, and routes that came with challenges many suburban stations never had to think about. The personalities reflected that intensity, both among associates and management. Every day demanded quick judgment, adaptability, and a willingness to deal with problems that did not have clean or comfortable solutions.

My focus in these stretch roles usually centered on three pressure points: Dispatch, RTS, and onboarding new DSPs. Dispatch alone could make or break an entire day. Vans backed up, routes weren't ready, devices failed, and drivers grew frustrated as minutes slipped away. Every delay compounded downstream. I spent most mornings identifying bottlenecks, reallocating labor, and making real-time decisions to get drivers loaded and out the door as close to on time as possible. In South-Central, that urgency mattered even more. Late departures often meant drivers delivering well into the night, increasing safety risks and burnout.

The DSPs operating out of that station were under constant strain. Many were dealing with high turnover, disgruntled former drivers, and real safety concerns that leadership outside the region rarely acknowledged. There were multiple incidents where Amazon vans were vandalized beyond repair, and in some cases, deliberately set on fire. These were not abstract risks. They were financial hits that DSP owners absorbed directly. Insurance claims dragged on, replacement vehicles were slow to arrive, and routes still needed to go out the next day regardless. Amazon's expectations never paused, even when a DSP had just lost a van overnight.

RTS was another battlefield. I worked closely with drivers and DSP leadership to reduce returned packages, knowing every return represented lost time, lost money, and a hit to their scorecards. We dug into root causes relentlessly. Access issues, unsafe delivery locations, poor routing logic, unrealistic stop counts, and outdated address data all fed the problem. Fixing it meant coordinating across teams, updating systems, and sometimes pushing back on assumptions that blamed drivers for failures rooted in bad planning. Progress was slow, but when returns dropped, you could feel the relief ripple through the operation.

Onboarding new DSPs in that environment was both rewarding and brutal. Many of these owners were first-time entrepreneurs who had bought into the promise of independence and growth. What they walked into instead was an unforgiving system with razor-thin margins and zero tolerance for missteps. I made it a point to be honest with them from day one. I walked them through Amazon's systems, yes, but I also talked about reality. Cash flow pressure. Route volatility. Labor challenges. The fact that one bad Peak could wipe out months of progress. My goal was not to sell them optimism. It was to give them a fighting chance.

Despite everything, most of these DSP owners showed up hungry to succeed. They cared about their teams, about safety, about doing the job right. Watching them navigate setbacks, absorb losses, and still push forward reminded me a lot of my own entrepreneurial past. I took pride in helping them find efficiencies, understand the scorecard, and avoid mistakes that could cost them their contracts. When a DSP stabilized, when a struggling owner found their footing, it felt like a real win in a system that didn't hand many out.

That stretch role in South-Central Los Angeles sharpened me in ways no single home station ever could. It deepened my operational expertise, but more importantly, it stripped away any illusions I still had about how unevenly risk was distributed in Amazon's Last Mile model. DSPs carried enormous responsibility with very little margin for error, while decisions made far above them could unravel their businesses overnight.

Looking back, that time reinforced something I still believe strongly. Logistics is not just about systems and metrics. It is about people absorbing pressure so the machine keeps moving. The relationships I built in those stations, especially with DSP owners fighting uphill battles every day, left a lasting impression on me. It reminded me that leadership, real leadership, shows up where things are hardest, listens when no one else is, and tries to make an unfair system just a little more survivable for the people stuck inside it.

Those stretch roles became a mirror I couldn't look away from. Every time I dropped into another station, especially South-Central, the same truth surfaced in a different form. Amazon's Last Mile machine depended on localized sacrifice. Risk was not shared evenly. It was concentrated in the places with the least margin, the fewest buffers, and the most exposure. DSPs in those environments were expected to perform at the same level as anyone else while absorbing dangers and losses that leadership would never experience firsthand. The dashboards flattened those realities into numbers, but standing on the floor, watching vans go out into neighborhoods where outcomes were unpredictable, made the imbalance impossible to ignore.

What stayed with me most was how often progress came from people compensating for a system that refused to adapt. DSP owners took hits they couldn't afford. Managers worked around broken assumptions. Drivers pushed through conditions that quietly increased risk with every late departure. And when things held together, Amazon took the win without examining what it cost. Stabilization was celebrated. The strain required to achieve it was forgotten.

By the time I rotated back to my home station after those assignments, something in me had shifted again. I was sharper, more capable, and more trusted than ever. But I was also carrying the weight of too many stories, too many near misses, too many examples of people holding the line so the machine didn't have to slow down. Those stretch roles confirmed what I had already begun to suspect. Amazon did not lack talented leaders or committed partners. It lacked a willingness to see how much damage its model asked others to absorb on its behalf. And once you see that clearly, it becomes harder to pretend the work is just about moving packages.

The Pandemic

Christmas Peak of 2019 at the Ashford station was unforgettable, the kind of stretch that burns itself into your memory. Every manager and associate was already operating beyond capacity, grinding through brutal hours as the holiday volume refused to let up. Peak is always chaos, but that year felt different. On top of the relentless surge of packages, an aggressive wave of flu-like symptoms ripped through the station, knocking out nearly every manager and a significant chunk of the workforce.

It wasn't a mild cold. Whatever it was tore through people fast and hard. Coughing fits, fevers, body aches, congestion that never cleared. The breakrooms looked like makeshift pharmacies. Mucinex, Advil, Tylenol Cold. People were taking them like Tic Tacs just to stay vertical. Most of us were clearly sick, but calling out wasn't an option. The volume didn't care, and neither did the deadlines.

As miserable as everyone felt, no one wanted to be the one who left their team hanging. There was a stubborn, almost reckless determination that took over. We showed up sick, exhausted, and barely functioning, because that's what the moment demanded. The shared misery created a strange kind of bond. We were all suffering together, dragging ourselves through twelve, fourteen, sometimes longer shifts with one goal in mind: get the packages out, get them delivered, make Christmas happen for customers.

In many ways, that stretch perfectly captured what working at Amazon felt like at its core. The mission came first. The deadlines were non-negotiable. Personal discomfort, health, even basic self-preservation took a back seat. No one stopped to question whether it was reasonable or sustainable. We just kept going.

When Peak finally broke, I took a few days off to recover, though "recover" is generous. Whatever I had lingered for weeks. The cough stuck around. The fatigue never fully lifted.

At the time, we chalked it up to a brutal flu season and the toll of Peak. It wasn't until months later, as news of COVID began spreading, that a different thought crept in.

Looking back now, I'm convinced we weren't dealing with just another seasonal bug. The symptoms, the speed at which it spread, the way it knocked out entire teams all at once. It lines up too closely to ignore. We were likely experiencing an early wave of COVID, long before it had a name or anyone in leadership was willing to acknowledge what was coming.

That realization is unsettling in hindsight. We powered through sickness without protections, without testing, without understanding what we were exposing ourselves to or carrying home to our families. And yet, at the time, it felt normal. Expected.

In a strange way, that Peak felt like a dress rehearsal for the storm that followed. It showed exactly how far Amazon would push, how much people would absorb, and how quickly "power through it" became the default response. We survived it, but it left a mark. One that didn't fully make sense until much later.

As 2020 unfolded, the world changed almost overnight, and Amazon's place in national logistics shifted with it. As brick-and-mortar retailers shut their doors or struggled to stay open, Amazon became a lifeline, delivering food, medicine, and basic necessities to millions of households. What had once been framed as convenience suddenly became critical infrastructure. At Ashford, the pressure didn't just increase, it exploded.

The stakes had never been higher. This wasn't about holiday gifts or Prime perks anymore. People were depending on us. That reality sat heavy on everyone in the building, and it pushed the operation into territory none of us had experienced before. That year, we broke records that hadn't even been thought possible. Ashford processed the highest package volume in Amazon's entire network. On our highest-volume day, we moved more than 200,000 packages, a number no other station had ever reached. And it wasn't a one-off. For days on end, we sustained volumes that should have broken the system.

The hours blurred together. Fatigue became background noise. People ran on muscle memory, caffeine, and sheer stubbornness. Processes were stretched, reworked, and sometimes held together by nothing more than experience and trust. Yet somehow, the operation kept moving. Vans rolled. Routes went out. Packages flowed.

What stood out most wasn't just the volume, but the mindset. Despite the relentless pace and constant pressure, no one folded. The team understood the moment we were in. This wasn't just another spike or another Peak. It felt bigger than Amazon. The work mattered in a way that was impossible to ignore, and that sense of purpose carried people through exhaustion that would have broken them under normal circumstances.

The challenges of 2020 were unlike anything we had faced before, and they left scars. But they also forged something durable. The Ashford station was tested at every level, operationally, physically, mentally. And we didn't collapse. We adapted. We endured. We proved, to ourselves more than anyone else, that we could withstand pressure most people would never see.

That year permanently changed how I viewed Amazon, leadership, and resilience. We were pushed to the very edge of what was sustainable and then forced beyond it. And for better or worse, we proved that the machine could keep running, regardless of the human cost required to do so.

In hindsight, 2020 was a turning point, not just for Amazon, but for everyone who lived through it from the inside. It showed what we were capable of when driven by shared purpose and an unwavering commitment to customers who suddenly depended on us for essentials. The pandemic reshaped the logistics network in ways no one could have predicted. Staffing shortages, supply chain breakdowns, and constant uncertainty forced us to adapt in real time, often without guidance or margin for error. Every layer of the operation was tested, and many days felt like controlled improvisation under pressure.

But beyond the operational challenges, the true cost of that year was far more personal.

During that time, my friend Samuel from the South-Central Los Angeles station lost his life to COVID. His death was devastating and cut through any illusion that this was only about volume, efficiency, or performance metrics. It was a stark reminder that behind every record-breaking day were real people absorbing real risk. We were labeled an essential workforce, but that didn't make us immune. We showed up, often exposed, often exhausted, and sometimes paid the ultimate price.

Samuel's passing made it impossible to separate the success of the operation from the human toll it demanded. It forced me to confront a truth that had been building for years: keeping the network running came at a cost that was rarely acknowledged and too often accepted as collateral damage. That loss stayed with me, not as a statistic, but as a reminder that resilience has limits, and no system should require people to sacrifice their lives to prove it works.

Samuel's death put a hard stop to any remaining abstraction. From that moment on, the language around resilience, sacrifice, and essential work stopped feeling neutral. It became personal. Records and throughput meant less when weighed against an empty chair, a voice that would never come back over the radio, a family left behind. The machine kept moving, as it always did, but something in me slowed down. I could no longer look at success without also seeing who paid for it. The cost was no longer theoretical or distant. It had a name.

What haunted me most was how quickly the system normalized that loss. There were moments of acknowledgment, brief pauses, carefully worded messages, and then the volume kept coming. That was the bargain Amazon had quietly made long before the pandemic. Adapt fast, absorb the shock, and do not stop. 2020 proved the model could survive almost anything. It also proved that survival was measured only by output, not by what it consumed along the way. We were praised for endurance, not protected from harm.

By the end of that year, I understood something I could not unlearn. Amazon did not just test resilience. It relied on it. It assumed people would keep showing up, even when exhausted, even when afraid, even when grieving. And for a long time, we did. But resilience without limits becomes exploitation, and purpose without protection becomes sacrifice. Samuel's loss drew a line that no metric could cross. It forced me to reckon with a question that would follow me from that point forward. How much human cost is acceptable for a system to prove it works, and who gets to decide when the price is too high.

Sent Where Everything Was Breaking

Amid all of this, stations across the network were buckling under the strain. Associate shortages became routine. Fatigue was constant. Operational problems that once would have been manageable now piled up faster than anyone could address them. In response, regional leadership decided to deploy experienced managers to stations that were actively unraveling, reinforcing operations where staffing and support had collapsed.

As it turned out, I was the one "volunteered. That assignment sent me straight back to South-Central Los Angeles, the very station that had just lost Samuel. The timing felt heavy. Returning there so soon after his death made the situation feel less like an operational decision and more like a moral test.

The irony wasn't lost on me. While I was preparing to step back into the daily chaos of an overburdened station, much of leadership remained insulated in-home offices, managing the crisis from a safe distance. The gap between those experiencing the reality on the floor and those reviewing it through dashboards and reports had never been wider. Metrics told one story. The human cost told another.

Still, I didn't hesitate. If there was a way, I could help stabilize the operation, if I could ease the load for a team already carrying grief on top of exhaustion, I was going to do it. Saying no didn't feel like an option.

Those months felt surreal. The world had shifted in ways that still didn't feel fully real. Los Angeles, a city defined by congestion and constant motion, was suddenly quiet. The freeways that once swallowed hours of my life were nearly empty. A commute that used to take two hours shrank to twenty minutes, an eerie convenience in the middle of a global crisis.

That drive became a strange daily reminder of how upside-down everything had become. While the roads were clear, the weight of what waited at the station was anything but.

When I arrived back at the South-Central Los Angeles station, the change was immediate and jarring. The place I remembered as loud, fast, and constantly in motion felt subdued, almost hollowed out. The energy was gone. Operations had been scaled back, headcount was thinner, and everything moved under the weight of pandemic restrictions. Even so, the station kept pushing forward. The people who were still there showed up every day, masked, exhausted, and understaffed, doing everything they could to hold the operation together. It was a stark reminder of how much pressure was bearing down on every layer of the system.

My primary mandate was clear: improve Delivery Completion Rate for the DSPs. What became clear just as quickly was how unrealistic that ask had become. Conversations with DSP owners painted a bleak picture. Labor shortages had gutted their operations. To stay afloat, many had no choice but to cut their workforce nearly in half. The math was brutal. Fewer drivers meant remaining Delivery Associates were carrying double routes, longer days, and heavier physical and mental strain. Vans went out overloaded. Drivers came back drained. Margins vanished, but expectations didn't.

In an effort to stabilize things, the station ran multiple hiring events, hoping to rebuild the DA pipeline. The result was sobering. No one showed up. Not a few candidates. None. It wasn't a failure of advertising or scheduling. People simply weren't willing to take on the risk, the hours, or the uncertainty. The labor pool had dried up, and no amount of corporate optimism could change that.

Every day felt unpredictable. Staffing fluctuated. Safety concerns were constant. One positive test could sideline an entire team overnight. DSPs were forced to make impossible decisions, choosing between burning out the drivers they had left or missing deliveries and absorbing the scorecard damage that followed. There was no winning, only damage control.

I spent most of my time working directly with DSP owners, trying to find ways to support their drivers while keeping routes moving. That meant adjusting expectations, reshuffling coverage, pushing for rescues where possible, and sometimes just listening while they vented frustration, they had nowhere else to put. It took creativity, flexibility, and a level of collaboration that didn't exist on paper but was necessary to survive.

We managed to get through it, but it wasn't because the system suddenly worked better. It was because people adapted out of necessity. That period stripped away any remaining illusions about resilience being limitless. The station didn't overcome the crisis so much as endure it, carried forward by people absorbing far more than they should have been asked to. It was a sobering lesson in just how fragile the operation became once the human buffer disappeared.

To be fair, this was also one of the rare moments where Amazon's upper leadership showed genuine empathy for station-level employees. Someone high up clearly understood the gravity of what frontline workers were facing, and health and safety quickly became a visible priority. Stations were consistently supplied with protective equipment: masks, hand sanitizer, Clorox disinfecting wipes, and latex-free or nitrile gloves. This wasn't a symbolic gesture or a one-time shipment meant to check a box. The supplies kept coming, reinforcing the sense that leadership was, at least in this moment, taking worker safety seriously.

Beyond the supplies, Amazon rolled out dedicated sanitizing crews that moved through the station multiple times per shift, spraying and wiping down high-touch surfaces. Door handles, desks, equipment, scanners, and shared workspaces were regularly disinfected. It made a real difference, not just in reducing risk, but in easing the constant anxiety that hung over every shift. Knowing that someone was actively focused on cleanliness and prevention mattered more than leadership probably realized.

Some of the changes were small but striking. Door desks, long neglected and usually grimy from years of nonstop use, were suddenly spotless. Areas that had once been ignored in the daily rush were cleaned thoroughly and consistently. For the first time, it felt like the physical environment was being treated with the same urgency as the operation itself.

That attention to safety and cleanliness didn't erase the pressure or exhaustion, but it did create a station that felt measurably safer during an incredibly uncertain time. In the middle of chaos and fear, those steps sent a clear message: at least for once, protecting the people doing the work was part of the mission, not an afterthought.

For me, the combination of a strengthened immune system and Amazon's sustained focus on health and safety made it feel like I had a real chance of getting through that period without getting sick again. More importantly, the company's response reflected a level of care that many of us hadn't expected, but absolutely noticed. For once, it wasn't only about keeping productivity afloat. Leadership had stepped up in a visible way to protect the people doing the work.

That sense of being taken seriously mattered. It created just enough stability to help us keep showing up during one of the most difficult stretches any of us had faced. Feeling protected, even imperfectly, made the grind survivable.

Looking back, that focus on safety wasn't just an operational decision. It was a rare acknowledgment of the human beings behind the packages. In the middle of chaos, Amazon showed that empathy and efficiency didn't have to be mutually exclusive. It was a lesson that stayed with me and shaped how I approached leadership long after that moment passed.

That deployment back to South-Central Los Angeles marked a strange inflection point for me. For the first time, I saw Amazon genuinely slow down one part of the machine in order to protect the people inside it. Not enough to remove the strain, not enough to make the work humane, but enough to acknowledge reality instead of denying it. After years of watching safety, burnout, and morale treated as secondary concerns, that shift stood out. It proved something important. Amazon was capable of empathy when the risk was undeniable and public. The question was whether it would ever choose to sustain that empathy once the immediate crisis passed.

What stayed with me wasn't just the PPE or the sanitizing crews. It was how quickly the operation stabilized once people felt even marginally protected. Fear didn't disappear, exhaustion didn't lift, but trust ticked up just enough to keep things moving. That experience reinforced something I had believed all along. People will carry extraordinary loads when they feel seen and safeguarded. They break when they feel expendable. The difference isn't complicated, but it is inconvenient for a system built on relentless output.

By the time my stretch role ended, I was changed again. Sharper, more grounded, and far less willing to accept the old narratives about resilience and sacrifice. I had seen what happened when the human buffer finally ran out, and I had seen how quickly things improved when leadership chose care over denial. That contrast was impossible to unsee. It raised a quiet but persistent question that followed me back to my home station. If Amazon could lead this way during crisis, why had it spent so many years choosing not to when the damage was slower, quieter, and easier to ignore.

Pandemic Promotion

The pandemic brought no shortage of challenges, but it also surfaced a few rare bright spots. One of them came when Paul, my leader at the time, went out of his way to recognize the work I had been doing. Without me asking, he took the initiative to put together a detailed Promo Doc on my behalf. It wasn't just a formality. He documented my contributions, the projects I had led, and the measurable impact I was having on both the team and the operation.

That kind of advocacy mattered. The document became a key piece in evaluating my readiness for promotion, but more than that, it felt like validation. In an environment where effort often went unnoticed and recognition was inconsistent at best, having a leader willing to step up and put his name behind my work meant a lot. It was a reminder that, even inside a system that could be punishing and impersonal, good leadership still made a difference.

Amazon's promotion system follows a rigid leveling structure that runs from entry-level roles all the way up to executive leadership, with each level carrying more responsibility, authority, and compensation. Levels 1 through 3 cover hourly, frontline roles. These are the associates working inside warehouses and fulfillment centers, positions that usually require little to no prior experience and focus almost entirely on execution.

Level 4 is where salaried leadership begins. This is typically the entry point for Area Managers and many college hires brought in through Amazon's university recruiting pipeline. From there, Level 5 marks the move into mid-level management. Roles at this level, such as Program Managers or Operations Managers, come with broader ownership, people leadership, and accountability for results, not just tasks.

Level 6 is where expectations escalate sharply. These roles require deep operational knowledge and the ability to manage complex teams, projects, or entire functional areas. By Level 7, leadership becomes less about day-to-day execution and more about strategy. Senior Managers and General Managers at this level oversee large teams, influence regional or network-wide decisions, and are expected to think several moves ahead.

Director-level roles begin at Level 8, where leaders are responsible for major business units or critical programs that span multiple organizations. Above that sits the executive tier. Vice Presidents at Level 9, Senior Vice Presidents at Level 10, and ultimately the CEO at Level 11 shape Amazon's long-term direction, global strategy, and innovation agenda.

Officially, movement between levels is driven by performance, leadership capability, and alignment with Amazon's leadership principles. In practice, at least from what I witnessed, advancement often depended just as much on who you knew and who was willing to sponsor you. There was an unmistakable "good old boys" dynamic at play, where relationships, visibility, and internal politics could outweigh actual results or competence.

With each step up, compensation shifted more heavily toward stock and bonuses, reinforcing the expectation that higher-level leaders weren't just managing operations, but shaping growth and long-term outcomes. The irony was hard to ignore. As the financial incentives increased, so did the distance from the day-to-day realities of the work, widening the gap between decision-makers and the people absorbing the consequences of those decisions on the floor.

Paul's leadership team reviewed the Promo Doc in detail and, after careful consideration, approved my promotion. It was a meaningful milestone in my career, bringing expanded responsibility, a modest increase in salary, and additional stock options. More than the title itself, what stayed with me was Paul's effort. He went out of his way to build the document and advocate for me, not because he had to, but because he believed in my work and my potential. That kind of support mattered. It reinforced my respect for him as a leader and strengthened my own commitment to mentorship, collaboration, and helping others move forward whenever I could.

At the same time, daily volume across Amazon remained exceptionally high. Delivery Associates began returning in larger numbers, driven by a clear understanding that their earnings were directly tied to how much they delivered. The newly implemented sanitation protocols also helped restore confidence. DAs felt safer on the road and inside the station, which made a tangible difference in morale and attendance.

The promotion arrived during a moment when advancement carried less certainty and more exposure. Increased responsibility came without the usual support structures, layered onto conditions already strained by the pandemic. What might have felt like progress under normal circumstances instead became a test of endurance and judgment under sustained pressure. The title changed, but the risks multiplied. Understanding how advancement functioned in that environment mattered before examining how expectations continued to rise even as stability remained out of reach.

The Pandemic Fallout

Leadership at Amazon consistently relied, from what I experienced, on Defects Per Million Opportunities, or DPMO, as a core quality metric and a foundational tool borrowed from Six Sigma to evaluate process efficiency and expose areas that need improvement. In a logistics environment where precision directly determines whether a package arrives on time or not, a high DPMO is a clear signal that something in the process is breaking down. Those breakdowns eventually show up as missed promises to customers and higher operational costs that ripple through the network. DPMO is calculated by dividing the number of observed defects by the total units processed multiplied by the number of potential defect opportunities per unit, then multiplying that result by one million. For example, if a station processes 10,000 packages, records 50 defects, and each package has five possible defect points, the DPMO would be calculated as (50 ÷ (10,000 × 5)) × 1,000,000, resulting in a DPMO of 1,000. This approach creates a scalable and standardized way to measure quality across vastly different operations.

In a logistics or Amazon station context, a single package often carries multiple defect opportunities. One package might be evaluated for correct scan, correct routing, correct containerization, proper load out, correct delivery address, and on time delivery. Each of these represents a separate opportunity for failure, and if any one of them goes wrong, it is counted as a defect against that specific opportunity. This framing matters because DPMO is designed to normalize quality across processes of varying complexity. A workflow with many handoffs and decision points naturally has more chances to fail than a simpler one. By accounting for defect opportunities per unit, DPMO adjusts for that complexity and allows quality to be compared more fairly across stations, shifts, or workflows. That fairness only holds if defect opportunities are clearly defined, consistently applied, and auditable. When definitions shift or counting is inconsistent,

DPMO can improve or worsen on paper without any real change in what is happening on the floor.

Within Amazon's operation, DPMO functions as a critical lens for evaluating accuracy and reliability across the entire logistics chain, from sort and route planning to final delivery execution. When tracked over time, it helps surface defect patterns, points teams toward root causes, and justifies corrective actions aimed at tightening execution. A lower DPMO reflects stronger process discipline and fewer errors, which ultimately translates into fewer delays, fewer misdelivered packages, and a delivery experience that feels predictable instead of chaotic to the customer.

Beyond being just another metric on a dashboard, DPMO functions as a roadmap for continuous improvement. A high RTS DPMO, for example, can expose breakdowns in return logistics that might otherwise be written off as noise. Left unaddressed, those issues quietly drive up costs, create delays, and erode the customer experience. By pushing DPMO down, Amazon is not simply chasing cleaner numbers. It appeared to be forcing attention onto the underlying process failures that generate rework, waste, and frustration across the network.

Reducing DPMO strengthens the efficiency of the logistics operation while simultaneously cutting the hidden costs tied to errors and exceptions. Just as importantly, it reinforces the company's promise of speed and reliability, which sits at the core of Amazon's customer obsession narrative. In that sense, improving DPMO is not only about quality assurance. It is about protecting Amazon's competitive position in a marketplace where even small execution failures are felt immediately by customers and magnified at scale.

At the South-Central Los Angeles station during the pandemic, RTS DPMO was in free fall. Missed deliveries were stacking up day after day, and the operation was struggling to keep pace with demand that never seemed to let up. Using data pulled from Mercury, the real time tool used to monitor delivery progress and package status, I leaned on my analytical background to break down on road DSP performance, with a narrow focus on undelivered packages. The data quickly showed where the operation was failing, but the numbers alone were not enough. To fix the problem, I needed to understand why those failures were happening in the first place.

I began generating detailed reports for each Delivery Service Provider (DSP), tracking metrics like successful delivery rates alongside specific failure reasons such as Unable to Access (UTA), Unable to Locate (UTL), Nonsecure Location, (NSL) and Business Closed (BC). Almost immediately, one anomaly jumped off the page. Business Closed cases were far higher than anything I had seen elsewhere. On the surface, the explanation seemed obvious. It was the pandemic, businesses were closed, end of story. But the data did not fully support that assumption. If widespread closures were the true driver, other stations should have been seeing the same spike. They were not. That inconsistency raised a red flag and pushed me to dig further instead of accepting the easy explanation.

To validate what the data was suggesting, I started contacting businesses that had been marked as closed in the delivery reports. What I found was telling. Most of them were open, just operating with reduced hours or skeleton crews. That discovery shifted the entire narrative. When I analyzed the timestamps tied to those Business Closed scans, a clear pattern emerged. Many of them occurred after standard business hours, well past when a legitimate delivery attempt should have been made. It became evident that some Delivery Associates (DA) were skipping delivery attempts during open hours and later marking packages as Business Closed to avoid the stop altogether. That behavior was quietly inflating RTS volume, driving up DPMO, and masking a solvable execution problem behind what looked like a pandemic driven inevitability.

Digging deeper, I cross referenced delivery timestamps against stated business hours and route sequencing, which made the pattern impossible to ignore. Certain Delivery Associates repeatedly showed up in the data as avoiding business deliveries altogether. This was no longer a vague trend. The behavior was traceable to specific routes and individuals. With that evidence in hand, I brought the findings directly to DSP owners, not as an accusation, but as a clear case for accountability and corrective action.

The impact was immediate and measurable. Once the behavior was addressed, RTS volume began to stabilize and the station's RTS DPMO dropped accordingly. More importantly, the operation regained predictability. Packages stopped cycling back unnecessarily, routes flowed more cleanly, and customer commitments were met more consistently. That experience reinforced a lesson I carried forward. Data, when paired with skepticism and follow through, has real power. There are many tools that can surface the truth, but only if someone is willing to question easy explanations, dig into the details, and push uncomfortable conversations in the service of fixing what is actually broken.

I made it a priority to attend the DSP's Stand-Up Meetings (SUMs) at their offsite parking lots, seizing the chance to connect directly with DAs in the field. These meetings became my time to inform, encourage, and motivate, especially as we navigated the early and uncertain days of the pandemic. I recognized that DAs were on the front lines of delivering essential goods, often to households under quarantine or lockdown, and they needed clear, practical guidance and reassurance.

At that point, I made the decision to start attending Stand Up Meetings for the lowest performing DSPs parking lots, and eventually expanded that effort to include all DSPs at the station. The goal was simple and intentional. I wanted to show visible support from Amazon leadership and create space for DAs to speak honestly about what they were dealing with on the road. These meetings were not meant to be another top-down lecture or a review of metrics. They were meant to give

DAs a place to vent, to call out what was not working, and to offer ideas for improvement, instead of being treated like anonymous data points disconnected from the reality of the job.

At each SUM, I made a point to walk through the latest safety protocols Amazon was rolling out and explain why they mattered. I framed those updates around protection, not compliance, emphasizing how these measures helped safeguard not only the associates themselves but also the families and communities they were delivering to every day. Beyond the official guidance, I shared practical, real-world tips that could make their day safer without slowing them down. Something as simple as using a package to press elevator buttons to avoid repeated contact with high touch surfaces was a small change, but one that reduced risk while keeping routes moving.

In those meetings, I spoke plainly and from the heart, and the response was immediate and visible. You could see it in their faces. For many of them, it was not about being told what to do. It was about feeling acknowledged and respected during a moment when their work carried more weight than ever. I wanted them to understand that they were not just delivering boxes. They were providing a lifeline to people who depended on Amazon's network to get through the pandemic safely. Every SUM became an opportunity to reinforce that message and to show them that their work mattered.

Through these efforts, my hope was not only to improve safety and efficiency, but to make a real difference in how DAs experienced their work. The gratitude I saw in the room mattered, but I knew the true impact would show up later on the road, in how confidently they ran their routes and how seriously they took both safety and execution. In a world defined by uncertainty, this was my small way of helping them feel supported, prepared, and proud of the critical role they played inside Amazon's operation.

As with any high demand job, there were always a handful of disgruntled Delivery Associates, most often frustrated by pay and long hours. The work was undeniably tough, especially during peak periods when routes stretched longer and the pressure to deliver more packages in less time intensified. Those complaints were not imagined or exaggerated. The workload was real. What often went unspoken, though, was that DSPs were operating under their own set of constraints. Amazon's compensation model paid DSPs a fixed rate per successfully completed route, leaving little flexibility to raise wages or layer on meaningful incentives. To stay afloat, DSPs had to run lean, which effectively capped what they could offer DAs regardless of how demanding the job became.

Inside some DSPs, that tension was impossible to miss. Frustration simmered beneath the surface and occasionally spilled over in alarming ways. At one offsite parking lot, those frustrations boiled over into a moment that no one could ignore. A delivery van was found burned out, widely believed to be an act of retaliation by a disgruntled DA. The message was unmistakable. It reflected the depth of anger some drivers felt and, at the same time, exposed the limited tools DSPs had to address those grievances. Caught between Amazon's relentless delivery expectations and a workforce stretched to its limits, DSPs were often left absorbing the fallout.

That charred van became a stark symbol of the strain running through the system. It forced a harder conversation about the imbalance between operational efficiency and human sustainability. It was no longer enough to push volume and tighten metrics. There was a growing need for Amazon and its DSP partners to rethink how DAs were supported, whether through better communication, recognition, flexibility, or creative ways to boost morale. Without that balance, the pressure only continued to build, and the cracks in the system became harder to ignore.

By the end of that stretch, DPMO no longer felt like a neutral quality metric. It felt like a mirror. When it spiked, it reflected not just process failures, but exhaustion, disengagement, and pressure that had nowhere left to go. The math could explain how defects accumulated, but it could not explain why people were making the choices they were making on the road. Those answers lived in missed breaks, unrealistic routes, capped pay, and a system that demanded precision without offering much margin for being human.

What unsettled me most was how often the response stopped at the number. When DPMO dropped, leadership celebrated. When it climbed, pressure increased. Rarely did the conversation linger on what the data was quietly pointing to underneath. Burnout. Fear. Disconnection. The burned-out van in that parking lot was not just vandalism. It was a flare. A visible expression of strain that dashboards were never designed to capture and leadership rarely wanted to confront directly.

Standing in those SUMs, talking to drivers who were risking their health to keep the network alive, I could feel how thin the line had become. People were still showing up, still delivering, still caring enough to argue about fairness and pay. But that caring came with volatility. When systems push hard enough for long enough, pressure doesn't disappear. It leaks. Sometimes into metrics. Sometimes into silence. Sometimes into fire.

That period crystallized another uncomfortable truth. Amazon's operational excellence depended on people constantly compensating for structural imbalance. Data could diagnose failures and help correct behavior, but it could not substitute for sustainability. You could drive DPMO down for a quarter or a Peak. You could stabilize RTS and clean up scorecards. But unless the human load was addressed, the system would keep generating the same failures in different forms. And once you see that clearly, it becomes impossible to believe that tighter metrics alone will ever be enough to hold it together.

Using Real-Time Data to Fix What Was Broken

The On Time Dispatch metric, or OTD, sat at the center of daily pressure inside the station, and tight scheduling was slowly strangling our ability to hit it. DSPs were given a rigid 30-minute window to load their vans and clear the building. That window dictated everything downstream, from route sequencing to how many hours Delivery Associates had to finish their routes. When vans missed that cutoff, the risk was immediate. Late dispatches cascaded into late deliveries, and once that happened, customers started missing their Promised Delivery Dates. The margin for error was almost nonexistent.

The South-Central Los Angeles station had a serious OTD problem. The issue was not effort. It was space and flow. Vans were stacked up outside the warehouse, waiting to access the load out area where their route racks had already been staged. The entrance quickly became congested, and once that happened, one delay rolled directly into the next wave. What should have been a tightly controlled process turned into a daily choke point that rippled through the entire operation.

To break that cycle, I pushed a solution that did not require more time, more labor, or additional headcount. Instead, I rethought how we used the space we already had. I created a secondary load out area in the large parking lot adjacent to the station and reorganized DSP positioning based on daily route volume, setting the foundation for a smoother, more predictable flow out of the building.

Smaller volume DSPs were reassigned to load out in the parking lot, while larger volume DSPs remained inside the warehouse, closer to where their racks were staged. Dispatch associates could pull the lighter racks out to vans in the parking lot and, because fewer racks were required, quickly return to support operations inside the warehouse. This reduced the distance and effort associated with moving loaded racks into the yard, cutting down unnecessary movement and speeding up overall turnaround time.

Keeping higher volume DSPs closer to their racks allowed them to load more efficiently and clear the building faster. DSPs that still could not complete load out within the dispatch window were redirected to the secondary location outside, where associates shuttled route racks out to them without disrupting the primary flow. The work did not slow down. It moved differently, and that shift in flow made all the difference.

Once the congestion cleared, the impact was immediate. Load out stopped bottlenecking, vans exited on time, and dispatch stabilized. OTD steadily climbed until it reached 98 percent. That improvement did more than clean up a metric. It restored predictability to the day, protected customer delivery promises, and removed a constant source of stress from both DSPs and station leadership. The fix was not complicated. It was simply a willingness to step back, question the physical layout, and adjust the system instead of blaming the people working inside it.

One of the most persistent challenges we faced was managing Out of Delivery Time (OODT) packages, which occurred when drivers simply ran out of time during their shifts and returned with undelivered packages still on their vans. This was especially common among newer drivers who were still learning their routes or struggling with navigation in unfamiliar areas. Layered on top of that was another costly issue: Pickup Failed packages. These were packages that missed critical scans during induction or sort and never populated in a driver's GPS. From the driver's perspective, they did not exist. As a result, they often went unnoticed until the end of the route, triggering unnecessary returns and adding friction back into the system.

The data models I built became a turning point in addressing both problems. They provided granular visibility into failure points, down to individual routes and even specific stops. With that level of detail, DSP dispatchers could flag missed packages before drivers ever left the station, instead of discovering them hours later at RTS. Drivers were able to deliver those packages in real time and notify dispatch, who in turn alerted the RTS team so the deliveries could be correctly closed out in the system. That feedback loop reduced OODT and Pickup Failed incidents, cut avoidable returns, and stabilized last mile execution. More importantly, it reinforced a broader lesson. When analytics are applied with intent and speed, they stop being retrospective scorekeeping and start becoming tools that actively improve how the operation runs.

Beyond fixing immediate delivery failures, those models became a powerful way to surface deeper, systemic issues across DSPs. By tracking OODT rates and other return categories over time, the data made it clear which DSPs were consistently struggling and where those struggles were concentrated. More importantly, it exposed patterns at the driver level that were invisible in high level rollups. That insight allowed DSP managers to move away from blanket feedback and generic coaching conversations that rarely changed behavior. Instead, they could sit down with specific drivers and point to concrete, data backed examples of where routes were breaking down and why.

That shift changed the tone and effectiveness of coaching. Conversations became more focused, more credible, and far more actionable. Drivers were no longer being told they just needed to move faster or try harder. They were being coached on specific stops, navigation issues, time management problems, or missed processes that were driving returns. Over time, that precision led to measurable improvements and more consistent performance. The real value of the analytics was not in exposing weak spots. It was in giving teams the clarity they needed to make smarter decisions, coach with confidence, and improve the operation without burning people out or guessing at the root cause.

To reinforce those data driven improvements, I made a point of showing up at DSP Stand Up Meetings and engaging directly with Delivery Associates. These sessions gave me a chance to move beyond dashboards and talk through practical, real world strategies for reducing returns in high impact categories like Unable to Access, Business Closed, and Out of Delivery Time. The goal was not to lecture or assign blame, but to create a space where drivers could understand the why behind the numbers and take ownership of problems before they snowballed into end of day failures.

Over time, those conversations helped shift the mindset from reactive cleanup to prevention. Drivers began thinking ahead about access issues, business hours, and time management instead of discovering problems at the end of the day. Paired with the analytics work, that hands on engagement produced measurable results. OODT dropped, Delivery Completion Rate improved, and more packages reached customers without cycling back through the building. Even with that progress, one major pain point remained. The RTS process itself was still inefficient, and fixing the breakdowns in debriefing quickly became my next priority.

The existing single lane debriefing setup had become a major bottleneck. Vans routinely spent more than 25 minutes stuck in line just to complete the return process before they could exit the station. That congestion backed traffic up into the yard, frustrated drivers who were already at the end of long shifts, and turned RTS into one of the slowest, most painful parts of the day.

To eliminate that choke point, I redesigned the process entirely by introducing a restructured RTS Pit Stop system. The single lane was replaced with three dedicated lanes: two lanes for vans returning ten or more packages, and one lane for vans returning fewer than ten or no packages at all. The impact was immediate. Debriefing time dropped to roughly 1.5 minutes per van, lines disappeared, traffic flow inside the station stabilized, and Delivery Associates were able to clear the site quickly and get back to their lots and home to their families instead of sitting in idling vans at the end of an already exhausting day.

A significant gap in the operation was the lack of real insight into the barriers Delivery Associates were facing on the road, which often resulted in packages cycling back to the station with little context. To address that blind spot, I introduced the RTS DA Questionnaire, a simple but effective tool designed to capture first hand feedback on undelivered packages. RTS associates would ask DAs a short set of targeted questions during debrief and record their responses, which were later aggregated and analyzed for patterns.

The questionnaire focused on common failure categories such as Unable to Access, Unable to Locate, Non Secure Location addresses, and other recurring delivery obstacles that rarely told the full story in system codes alone. These brief but intentional conversations allowed us to document what actually happened on the route and share that information with the Field Quality Assurance team so issues could be addressed upstream rather than repeatedly absorbed at RTS.

The initiative was well received by DAs, many of whom appreciated being asked for their perspective and included in the problem-solving process. By creating a consistent feedback loop, we were able to refine strategies using real, on the ground insight instead of assumptions. Over time, as the data was analyzed and targeted adjustments were made, RTS returns steadily declined. It became a clear example of how pairing structured data with direct input from the field could drive meaningful and measurable improvements in last mile execution.

The success of the RTS DA Questionnaire did not go unnoticed, but it was quickly tangled up in a familiar Amazon reality. A leader saw the results, repackaged the idea, and presented it as their own in pursuit of a promotion. It was not an isolated incident. It was part of a broader cultural flaw where individual ambition often outweighed collaboration. The expectation on paper was that ideas should be shared freely for the good of the operation. In practice, credit for meaningful work too often flowed upward to whoever needed a win, regardless of who actually built the solution.

As frustrating as that was, it did not erase the impact of the work. RTS returns dropped, Delivery Associates finally had a voice in shaping operational improvements, and the station benefited from cleaner, more efficient processes. The results spoke for themselves. In the end, the real measure of success was not a promotion or a line on a resume. It was the fact that drivers spent less time stuck in debrief, fewer packages cycled back unnecessarily, and the operation ran with less friction. That was the outcome that mattered, whether or not the credit landed in the right place.

That approach reflected how I used data throughout my time at Amazon. I built a series of models, mostly in Excel, with the specific intent of exposing root causes rather than surface level symptoms. Data was pulled from multiple sources, cleaned, aggregated, and shaped through pivot tables and visualizations until patterns emerged that would have been easy to miss in raw reports.

The goal was never to point out inefficiencies for the sake of reporting. It was to fix them while the operation was still moving. Real time data only matters if it drives real time decisions, and those decisions only matter if they result in practical changes on the floor. That mindset consistently turned numbers into action and underperforming processes into measurable, repeatable gains.

By that stage, the pattern was impossible to ignore. Every meaningful improvement came from questioning assumptions, redesigning flow, and listening to the people closest to the work. None of the fixes required heroic effort or brute force. They required clarity. Space instead of congestion. Feedback instead of silence. Data used as a flashlight instead of a hammer. When those elements came together, metrics followed naturally. When they did not, leadership defaulted to pressure and blame, as if intensity alone could compensate for structural flaws.

What stayed with me most was how often the system confused discipline with insight. Amazon loved metrics, but it struggled with interpretation. Numbers were treated as verdicts instead of signals, and when leaders skipped the harder work of understanding root causes, they inevitably landed on the wrong conclusions. Drivers were blamed for not delivering due to bad routing. DSPs were blamed for poor delivery performance of their Drivers. broken. Stations were blamed for process decisions made far above them. The operation moved forward, but the same problems resurfaced again and again under new names.

That chapter reinforced something fundamental about how I worked and why it mattered. I was never interested in winning arguments or claiming credit. I wanted the floor to run better tomorrow than it did today. I wanted fewer drivers sitting in line after midnight, fewer packages cycling back for no reason, and fewer people absorbing stress that could have been designed out of the system entirely. The fact that those improvements often went unrecognized said less about their value and more about the culture they lived inside.

In the end, that work became its own quiet record. Cleaner dispatches. Faster RTS. Lower returns. Better conversations. Less friction. The machine ran smoother because people were finally being heard and systems were finally being questioned. And while Amazon rarely paused to acknowledge that kind of progress, I knew what it took to make it happen. That knowledge would matter later, when the question was no longer how to fix the operation, but whether staying inside it was worth the cost of continually doing so without protection, credit, or relief.

New Station Launches

Opening a new distribution center at Amazon is never a casual decision or a simple build. It starts long before any physical movement, with layers of market analysis meant to predict demand, shipping volume, and delivery speed down to the zip code. Leadership looks at population density, transportation corridors, labor availability, and how quickly packages can move in and out of the region. Zoning rules, local politics, and long-term growth projections all factor in. By the time a site is selected, the decision has already been debated, modeled, revised, and pressure tested more times than most people realize.

Once a location is locked in, the design phase kicks off, and this is where Amazon's obsession with efficiency becomes visible. Architects and engineers work hand in hand with operations teams to shape the building around flow, not comfort. Storage layout, dock placement, conveyance paths, and automation are all designed to reduce seconds and steps. Every square foot has a purpose. The goal is not just to house inventory, but to move it relentlessly, with as little friction as possible, while still meeting safety and throughput targets that will be enforced from day one.

Regulatory compliance runs parallel to all of this. Permits, environmental reviews, and local approvals are secured, sometimes smoothly and sometimes through drawn-out negotiations. Only after those hurdles are cleared does construction or refurbishment begin. Whether it is a ground-up build or a retrofit of an existing warehouse, the project is tightly managed. Timelines matter. Safety standards matter. Deviations from the plan are scrutinized because delays ripple downstream into launch dates, hiring schedules, and regional capacity commitments.

As the physical space comes together, the technology stack follows. Equipment is installed, warehouse management systems are configured, and automation is tuned to fit the specific layout of the building. Nothing runs on instinct. Everything runs on data, sensors, and rules that dictate how packages are sorted, staged, and pushed toward the next mile. At the same time, staffing ramps up. New hires are brought in, trained on safety, systems, and workflow expectations, and quickly introduced to the reality that this environment moves fast and does not slow down for learning curves.

Before the building goes live, trial runs expose weak points. Bottlenecks surface. Processes break. Adjustments are made, sometimes hastily, sometimes thoughtfully, depending on leadership. Only when the operation is deemed stable enough does the launch happen, followed by constant monitoring and relentless performance tracking. On the outside, Amazon often emphasizes community engagement and economic impact. Inside, the focus remains fixed on metrics, volume, and execution. The entire process reflects Amazon's larger philosophy. Build fast, optimize continuously, and expect the operation, and the people inside it, to absorb whatever pressure is required to meet customer expectations.

By this point, leadership had started to see me as a domain expert in Dispatch and RTS operation processes, someone who understood not just the metrics but how the work actually unfolded on the ground. That reputation is what put me on the launch team. I was assigned to the Eastgate Distribution Center, stepping into a role that blended training, stabilization, and expectation setting from day one. I trained Dispatch Yard Marshalls, helped onboard new DSPs and their delivery associates, and carried over the RTS Pit Stop system I had built and refined at the South-Central Los Angeles station.

The work was less about introducing brand-new ideas and more about establishing discipline early, before bad habits had a chance to harden into standard practice. Launch environments are unforgiving. Systems are new, people are inexperienced, and volume ramps faster than anyone would like. What I brought to Eastgate was a proven playbook and the credibility to enforce it. That assignment reinforced something leadership already believed about me. I could walk into a new or unstable operation, translate theory into execution, and create structure where there was none.

I took the lead on onboarding ten new DSPs and their delivery associates, using training materials I had built and refined through prior launches and recovery efforts. The objective was not to overwhelm new teams with theory, but to ground them in how the operation would actually function once volume ramped and pressure set in. The training covered core operational procedures, safety expectations, and practical efficiency habits that tend to separate stable DSPs from those that struggle early. I worked closely with DSP managers and their teams on the floor, tailoring guidance to their specific gaps and making sure they walked away with more than just documentation. They left with context, expectations, and a clearer understanding of what success would realistically require.

In parallel, I implemented the RTS Pit Stop process I had successfully deployed at other stations, building it directly into Eastgate's launch framework rather than trying to retrofit it later. Returns are where many new stations stumble as volume increases, so the focus was on designing a clean, predictable closeout from day one. The rollout required tight coordination with station leadership, DSP owners, and associates to align on flow, roles, and timing before bad habits could form. Instead of reacting to congestion after it appeared, the Pit Stop model established structure upfront, giving the station a scalable RTS process that could handle growth without turning nightly debriefs into a bottleneck as volume increased.

Because the groundwork was solid and expectations were set early, the Eastgate launch avoided many of the disruptions that usually plague new stations. Planning was deliberate, communication was tight, and proven processes were embedded before volume had a chance to expose weak spots. Eastgate now pushes tens of thousands of packages out the door every day, delivering what Amazon likes to call "smiles" to customers, but behind that slogan is an operation that was built to scale instead of scramble. For me, the launch reinforced a simple truth. When experience is paired with preparation and people actually work together, operational excellence is not an accident. It is engineered.

Not long after wrapping up work at the Eastgate Los Angeles station, I was handed the next challenge. I was asked to help open a brand-new Mega station in Mission Viejo, California. It was the kind of opportunity that does not come lightly and it marked a clear inflection point in my career. This was not about stabilizing an existing operation. This was about building one from nothing. As I prepared to step into that role, I felt both the excitement and the weight that came with it. I knew the station's success would hinge on leadership, disciplined execution, and the ability to translate Amazon's standards into something that actually worked on the ground.

Standing on the edge of that Rancho Sereno launch, I felt the familiar pull of responsibility settle back in. New buildings always came with a sense of possibility, the rare chance to do things right before shortcuts, bad habits, and political noise crept in. But they also came with risk. A Mega station magnified everything. Volume, expectations, scrutiny, and pressure all scaled faster than headcount or patience ever did. I knew from experience that the launch itself would be judged as a success or failure long before the people inside it had a real chance to find their footing.

What weighed on me most was how much depended on decisions made before the first van ever rolled out. Flow, layout, training, and culture were either set with intention or left to harden under stress. I had seen too many stations launched on optimism and dashboards alone, only to spend years paying for early missteps. This time, there was no illusion about what was at stake. If the foundation was weak, the operation would bleed quietly and endlessly. If it was strong, it could absorb growth without breaking the people inside it.

As I prepared to step into Mission Viejo, I carried everything that had come before with me. The wins, the mistakes, the people I had helped, the ones I had lost, and the lessons learned the hard way. I knew how to build a station that could move volume. The harder question was whether it could be built in a way that did not repeat the same patterns of burnout, disposability, and silent damage I had watched unfold elsewhere. That question lingered in the background as the launch approached.

At the time, I still believed the answer could be yes. That with enough experience, preparation, and care, it was possible to engineer not just operational excellence, but something more sustainable. Rancho Sereno felt like a proving ground, not just for the station, but for everything I had learned about leadership, systems, and the human cost of scale. I did not yet know how sharply that belief would be tested.

Mega Station Launch

Launching a Mega station is not an abstract exercise in planning. It succeeds or fails in the delivery process, and that is where my focus stayed at Mission Viejo. From the moment I arrived, leadership entrusted me with a pivotal task: building out the operational infrastructure that would allow the station to launch cleanly and function under real delivery pressure. That meant everything tied to dispatch, load out, RTS, and DSP execution, but it also meant the physical environment those processes depended on. A building without the right structure quickly becomes chaos once vans start moving.

Drawing on experience from Eastgate, I moved fast to assess what the station actually needed to support delivery flow. Signage, dividers, bollards, DSP desk markers, yard controls, and even basics like the Manager's Closet all had to be sourced and installed before volume hit. Armed with my Amazon corporate credit card and a clear mental map of how delivery operations break down when details are missed, I began filling the gaps. Every purchase was deliberate. Clear navigation paths for drivers. Defined staging and walking lanes. Organized, functional spaces for DSP leadership to operate without tripping over one another. My credit card definitely got a workout, but the goal was never optics. It was control, safety, and execution.

At the same time, I worked closely with station leadership, central operation teams, and DSP owners to design how delivery would function day to day. Yard flow, staging logic, wave timing, and RTS closeout were built with scale in mind, knowing that even small inefficiencies become painful once hundreds of vans are cycling through the yard. Equipment placement, technology setup, and space planning were all evaluated through one question. Does this make delivery easier or harder at scale? Training followed the same philosophy. Dispatch Yard Marshals, DSP managers, and delivery associates were coached not just on what to do, but on why the process mattered and how shortcuts would show up later in missed waves and broken days.

As opening day approached, my attention stayed on the floor and in the yard, where theory collides with reality. Welcoming new associates and DSP teams was less about ceremony and more about readiness. Rancho Sereno was going to be my home station, and I treated it that way from the start. Expectations around safety, timing, and accountability were set early, but so was space for feedback as the operation found its footing. Delivery operations are unforgiving. Weather, traffic, equipment failures, and fatigue all test the system daily. My job was to make sure the structure could bend without snapping.

The months after launch became a cycle of observation, adjustment, and coaching. Dispatch timing tightened. RTS flow stabilized. DSP onboarding matured. Weak points surfaced and were addressed before they turned into habits. Success was not measured solely by metrics, but by whether drivers could load efficiently, complete routes safely, and return to station without nightly pileups. The physical infrastructure we put in place proved just as important as the processes themselves, especially in high-traffic areas where confusion usually breeds delay and risk.

Rancho Sereno quickly became a proving ground for delivery operations at scale. The station spanned over 300,000 square feet on nearly 27 acres, designed as a megacycle hub serving Southern Orange County. Dual drive lanes flanking a central sorting and staging area allowed for high throughput, but only with precise coordination between dispatch and RTS. On paper, it was a showcase of Amazon's last-mile ambition. On the ground, it demanded disciplined delivery leadership to function. While the foundation we built supported early success, the station would eventually become one of the most challenging chapters of my career, not because the delivery processes failed, but because leadership did. What followed exposed how quickly even a well-designed operation can unravel when decision making, accountability, and support break down above the floor. It revealed just how much strain Amazon's delivery model places on the people expected to keep it moving every day, especially when leadership priorities shift away from execution and toward optics.

Eight of the DSPs I had previously onboarded transferred over from the Ashford Distribution Center, a transition that felt bittersweet. Those partnerships marked some of my earliest experiences working inside Amazon stations, and seeing that chapter slowly phase out carried more weight than I expected. Alongside supporting those familiar DSPs, I was also responsible for bringing several brand-new providers into the operation, making sure they integrated cleanly into a station that was still finding its rhythm.

I spent time walking the new DSP owners and their teams through the building, grounding them in how Rancho Sereno actually functioned. We focused on the load-out lanes, RTS flow, and the safety expectations that mattered most once volume and traffic picked up. Those early days were defined by hands-on guidance and constant conversation. Their energy and willingness to learn made the process productive, but it also reinforced how much early clarity matters for DSPs trying to get their footing in an unforgiving system.

The DSP owners were optimistic about the new station, particularly the expanded indoor Dispatch and RTS areas. That space proved invaluable, especially during bad weather, reducing stress on drivers and keeping operations from grinding to a halt when conditions turned. More than anything, those early interactions highlighted the value of trust and partnership. Supporting both experienced and first-time DSPs through the transition helped establish stability early on, laying a foundation for Mission Viejo's initial success and reinforcing how critical strong relationships are in keeping last-mile operations moving.

Before the official launch, I organized an open tour of the station for all participating DSPs, giving them a behind-the-scenes look at how Rancho Sereno was set up to operate. The goal was transparency and familiarity. We walked through the launch pads where vans would stage for Dispatch and RTS, the sort and staging zones, key safety features, leadership offices, and the areas designated for Delivery Associate training. Leadership led the tour and explained each space in detail. I stayed deliberately toward the back of the group, not disengaged, but watching closely. I wanted to make sure no one was left out of the conversation and to get a feel for how the DSPs were interacting with one another.

There was already tension in the air. A sensitive issue had been circulating among the DSPs before the tour even began. Several owners believed other DSPs were poaching their drivers. In a system where hiring and training a DA costs real time and real money, those accusations carried weight. Standing at the back gave me a clear view of side conversations, body language, and the unspoken frustration that leadership either did not see or chose not to acknowledge. It was obvious this tour was doing more than orienting people to the building. It was exposing fault lines within the DSP community that would eventually have to be addressed.

As we moved through the station, that tension finally broke the surface. What was meant to be an informational walk-through was overtaken by a heated exchange between two DSP owners. One openly accused the other of trying to lure away several of his drivers. Voices rose. The mood shifted instantly. Then it escalated fast. One DSP, visibly angry, threatened to "beat the shit" out of the other. Before anyone could fully react, his son, a large and imposing presence, stepped forward and shoved the accused DSP. In that moment, the tour stopped being about operations and became a real safety concern.

What stood out to me was how quickly a professional setting unraveled when underlying issues were ignored. This was not about a lack of space, signage, or process. It was about pressure, distrust, and competition inside a model that pits DSPs against each other while expecting collaboration. The incident reinforced something I had already learned through experience. You can build a clean station, design efficient delivery flow, and plan every detail, but if leadership fails to address the human and relational strain baked into the system, it will surface eventually, and it rarely does so quietly.

Acting on instinct, and probably recklessly, I stepped between them and used everything I had to keep the situation from turning physical. Leadership was only a few steps ahead of us and either did not notice what was happening or chose not to intervene. Either way, the responsibility landed squarely on me. What made it even more surreal was who they were. Both DSPs were friends of mine and seasoned professionals. One was a former Marine MP. The other had flown Black Hawk helicopters for the Army. Watching that level of discipline and shared history collapse in the middle of a station tour was unsettling in a way that stuck with me.

I leaned on that shared respect and spoke to them not as leaders inside Amazon's system, but as people who understood consequences. Somehow, the words landed. After a tense standoff, they backed down and agreed to stop the poaching. The moment passed, but the weight of it lingered. It was a sharp reminder of how fragile professionalism becomes when pressure, money, and survival collide inside a model that quietly encourages competition while preaching partnership.

The fallout was awkward and unavoidable. Their relationship stayed strained throughout the first year, with every interaction carrying a trace of what had happened. Over time, though, shared challenges began to soften the edges. Volume, breakdowns, staffing gaps, and the daily grind forced cooperation. Slowly, the tension eased. One DSP, who owned an auto repair shop, began helping the other with vehicle maintenance. The other returned the favor by lending extra vans when capacity got tight. What nearly turned violent eventually evolved into something functional, even supportive.

That arc stayed with me. It showed how much resilience exists among DSPs when they are pushed to the brink and still choose cooperation over resentment. It also reinforced a harder truth. The system creates these fractures, then steps back and lets people absorb the fallout. Unity, when it happens, is rarely because of leadership. It happens despite it, forged in moments where people decide that surviving together is better than tearing each other apart.

That moment at Rancho Sereno stayed with me because it stripped the illusion away in real time. We had done everything right on paper. The building was designed to scale. The processes were tight. The delivery flow worked. And still, the first real fracture came not from volume or technology, but from people being pushed into competition inside a model that quietly depends on their cooperation. Leadership talked about partnership, but the incentives told a different story. When pressure rises and margins thin, trust becomes the first casualty.

What struck me most was how little protection there was when things went sideways. The conflict did not de-escalate because of policy or leadership presence. It de-escalated because individuals stepped in, absorbed risk, and relied on shared respect forged long before Amazon entered the picture. That was a pattern I had seen repeatedly. The system created tension, then relied on human restraint to keep it from exploding. When it worked, Amazon took the win. When it didn't, the fallout landed on everyone else.

Rancho Sereno made one thing painfully clear. Operational excellence can be engineered, but stability cannot be mandated. You can design flow, enforce metrics, and build infrastructure that hums under load, yet still fail the moment leadership refuses to engage with the human strain baked into the model. The DSPs eventually found their footing not because the system guided them there, but because survival demanded cooperation. That cooperation was fragile, earned the hard way, and entirely unsupported from above.

By then, I understood the role I kept being placed in. I wasn't just building stations or fixing processes. I was standing in the gap between a system that optimized relentlessly and people who absorbed the consequences of that optimization every day. Rancho Sereno would go on to test me further, but that early incident crystallized the truth. You can build a world-class operation, but if leadership refuses to own the pressure it creates, the cost will always surface somewhere. And more often than not, it surfaces in moments where someone has to step forward and stop things from breaking in ways no metric can capture.

How Amazon Destroys Its Best People

Another manager and I were handed a deceptively simple but critical assignment: build the drive lanes that would control vehicle flow and safety inside the Rancho Sereno station. There was no template handed down and no margin for guessing wrong. We were responsible for designing it from scratch and making sure it worked once hundreds of vans started moving through the space every day. Armed with industrial-grade six mil vinyl tape with heavy rubber adhesive, we treated the project less like a facilities task and more like an operational control system.

We started outside the building, mapping out staging zones where vans would queue before entering the station. Every spot was measured deliberately. Too tight and drivers would stack up. Too loose and valuable yard space would be wasted. We taped each holding area directly onto the concrete with peak volume in mind, knowing that what felt spacious on an empty day would feel claustrophobic once waves began backing up. The goal was controlled flow, not just order. Vans needed to move predictably and safely, even when pressure was high.

Once the exterior staging was complete, we moved inside and laid down the internal drive lanes. These markings guided vans from entry to their designated parking spots and had to align perfectly with dispatch timing, pedestrian walk paths, and safety clearances. Every line mattered. A few inches off could create blind spots, pinch points, or unnecessary risk. We chose the heavy-duty tape specifically because we knew it had to survive constant traffic without peeling or fading. This was infrastructure that needed to hold up under abuse, not just look clean on day one.

The work was physical and methodical, but the intent went far beyond laying tape on concrete. We were building structure into an environment that would soon be under constant strain. Clear lanes reduced confusion, improved safety, and removed hesitation from drivers who were already juggling tight schedules and unfamiliar surroundings. Those markings became the backbone of smoother Dispatch and RTS flow. Seeing vans move cleanly through the lanes during early operations was one of those quiet wins that rarely gets recognized but makes everything else possible. It reinforced something I had learned over and over. In high-pressure environments, precision is not cosmetic. It is what keeps the operation from breaking when it matters most.

The emotional strain on leadership became impossible to ignore on D-1, the day before launch. As the countdown closed in, station leaders who had spent months preparing began second-guessing decisions that were already settled. What followed was a flood of last-minute changes that ignored operational reality and introduced confusion where clarity was most needed. Instead of trusting the plan, leadership reacted to anxiety, and that reaction exposed a deeper flaw I had seen before inside Amazon. Too often, leaders are placed into high-pressure launch environments without the operational depth or business maturity required to stay steady when it matters most. The result is reactive decision-making that creates risk at precisely the moment the system needs discipline and restraint.

What should have been a long but manageable thirteen-hour day turned into a twenty-two-hour grind as we scrambled to unwind and rework changes that never should have been made. By the time launch day arrived, everything was technically in place, but the team was running on fumes. Exhaustion hung over the operation before the first van even rolled in. Predictably, day one did not go as planned. The most immediate breakdown came from DSPs missing their assigned Wave Times, the tightly controlled thirty-minute windows designed to bring drivers in, load vans, and clear the yard in sequence.

Mission Viejo's layout only magnified the problem. With both North and South drive lanes in play, DSPs entered through the wrong lanes or arrived outside their assigned windows, clogging flow and creating bottlenecks almost immediately. What had been carefully designed to move smoothly unraveled under confusion and fatigue. It was not a systems failure. It was a leadership failure, rooted in poor communication, last-minute indecision, and a lack of confidence in the very processes that had been built to support the launch.

That chaotic start reinforced a lesson I would relearn repeatedly. Preparation alone is not enough. In complex operations, especially at launch, success depends on leaders who can execute under pressure with clarity, trust their teams, and resist the urge to micromanage when anxiety spikes. Without that steadiness, even the most well-designed systems can be undermined before they ever have a chance to prove themselves.

For the first few days, the Wave Time process threatened to unravel the entire operation. Vans arrived early, late, or in the wrong lanes, and each miss compounded the confusion. The turning point came from one of my Shift Assistants, Ed, who saw the problem for what it was. Not a lack of rules, but a lack of clarity. His solution was almost embarrassingly simple. He created laminated cards that listed each DSP's assigned Wave Time for both the North and South drive lanes. The cards clipped directly onto the DAs' ID badge lanyards, putting the information exactly where it was needed.

The impact was immediate. Drivers began showing up at the correct lanes, at the correct times. Dispatch stabilized almost overnight. What dashboards and reminders could not fix; a piece of laminated paper did. It was a reminder that the best operational solutions are often the ones that meet people where they actually are, not where leadership assumes they should be.

Ed continued to distinguish himself in the months that followed. He consistently brought practical, creative solutions to problems that others overcomplicated. One of his most effective ideas was creating a contingency area near the exit doors for drivers who could not finish loading within their Wave Time. Instead of blocking lanes and backing up the operation, those drivers could pull aside, complete their load, and rejoin the flow without disruption. The improvement was immediate and measurable. The process worked so well that it was eventually adopted as a standard practice across Amazon's station network. Ed's contributions reinforced something I believed strongly. Real innovation in operations rarely comes from policy decks. It comes from people close enough to the work to see what actually needs fixing.

Beyond the tangible fixes he delivered, Ed embodied Amazon's Leadership Principle "Have Backbone; Disagree and Commit" in the way it was supposed to look in practice. He was not afraid to challenge broken processes, even when doing so meant pushing back against assumptions that leadership had already accepted. He did it respectfully, with data and logic, and more importantly, with alternatives that actually worked. Ed did not complain for the sake of complaining. He identified problems, proposed solutions, and then committed fully once a direction was set. That combination of courage, creativity, and follow-through made him one of the most valuable people on the floor and a quiet driver of the station's early success.

Because of that, I pushed hard for his promotion during Performance Level Review meetings. On paper and in reality, he checked every box. Strong judgment. Operational impact. Leadership presence. Peer respect. Despite that, the recommendation was rejected. It was a deeply frustrating outcome, but not a surprising one. It reflected a recurring flaw I saw across Amazon's leadership pipeline, where inexperienced or disconnected leaders failed to recognize high performers unless they fit a narrow, often superficial mold. Decisions like that do real damage. They demotivate the very people who carry operations through chaos and signal that results matter less than optics or tenure. Ed's situation was not just a missed promotion. It was a clear example of how short-sighted leadership decisions quietly erode morale and undermine long-term operational strength.

Ed's willingness to challenge the status quo, while invaluable to the operation, eventually put him on a collision course with station leadership. One flashpoint came during a discussion on Amazon's real-time communication platform app, where a leader proposed cordoning off part of the South Drive Lane to allow associates to walk through during active dispatch. Ed immediately pushed back. His concerns were grounded in reality. The idea introduced serious safety risks and would have disrupted vehicle flow at the exact moment precision mattered most. From an operational standpoint, the proposal made no sense. From a safety standpoint, it was reckless. Still, being right did not matter. His objection was not received as insight. It was taken as defiance.

What followed exposed a darker side of leadership behavior. My manager, Mark, attempted to retaliate by pressuring me to submit a negative report against Ed through ADAPT, Amazon's disciplinary system. The intent was obvious. It was not about correcting behavior or addressing performance. It was about undermining Ed's credibility for speaking up. I refused. To me, it was a clear misuse of authority and a direct violation of the principle's leadership was supposed to uphold. ADAPT was being weaponized to silence dissent, not to protect the operation.

That refusal marked a turning point. Tension between Mark and me escalated immediately. The irony was hard to ignore. I had trained Mark myself, yet here I was being asked to compromise my integrity to protect his ego. I would not do it. That moment fundamentally shifted our working relationship and reinforced something I was already beginning to understand. Inside Amazon, challenging bad ideas can carry consequences, even when those challenges are rooted in safety, experience, and doing the right thing.

This situation exposed the darker side of Amazon's leadership culture, where ethical considerations were often sacrificed for personal agendas or political maneuvering. Instead of fostering collaboration and trust, leaders like Mark weaponized systems meant for accountability to settle scores and assert dominance. Navigating these dynamics was a constant challenge, forcing me to advocate for integrity in an environment that too often rewarded compliance and self-preservation over doing what was right.

Not long after that conflict, Ed was dismissed without severance. It was a staggering loss of talent and institutional knowledge, handled in a way that made it clear the decision had far more to do with internal politics than performance. Someone who had helped stabilize dispatch, improve safety, and introduce processes that were later adopted network-wide was gone in an instant. There was no acknowledgment of his contributions, no attempt to course-correct, just a clean removal that reflected how disposable high performers can become when they make the wrong people uncomfortable.

In the days that followed, one of the DSP owners pulled me aside and asked a simple question. Would I recommend Ed as a Delivery Associate? I did not hesitate. I told him exactly who Ed was, how he worked, and why losing him was Amazon's mistake, not Ed's. The DSP hired him shortly after. Since then, Ed has thrived. He earned multiple promotions within that organization and continued doing what he always did best, solving problems, leading from the front, and elevating the people around him. Watching him succeed outside Amazon only reinforced what I already knew. Talent does not disappear when a system rejects it. It just finds a place where it is allowed to grow.

That episode became a stark reminder of how damaging flawed leadership decisions can be. Amazon's failure to recognize and retain someone as capable and innovative as Ed did real harm, not just to morale, but to the operation itself. It exposed a broader pattern where internal politics and short-term optics routinely outweigh performance and results. In a culture like that, losing someone who consistently brought ingenuity, discipline, and a genuine commitment to doing the job right is not just unfortunate. It is self-inflicted damage.

Ed's departure stands as a cautionary tale. When leaders punish people for speaking up, for challenging unsafe or inefficient ideas, they do more than silence dissent. They drain the organization of the very people who keep it functional under pressure. By failing to value creative problem-solvers and principled operators, Amazon weakens the foundation it depends on to scale and execute. This was not about one individual leaving a station. It was about a system that repeatedly chooses comfort over courage and politics over merit.

In the end, the lesson was hard to ignore. Organizations do not lose talent by accident. They push it out through decisions that reward compliance instead of integrity and ego instead of results. Ed succeeded the moment he landed somewhere that valued what he brought to the table. Amazon, meanwhile, lost someone who made the machine work when it mattered. That imbalance says far more about leadership culture than any leadership principle ever could.

When Ed left, I naively believed that the worst of Amazon's leadership culture had left with him. I was wrong. Not long after his departure, another station leader pulled me aside and dropped a demand so absurd it barely registered at first. He wanted me to figure out a way to terminate Ahmad's employment. There was no buildup, no context, just the expectation that I would make it happen.

My first reaction was disbelief. I asked the obvious question. Was there a performance issue? A safety concern? Anything concrete that would justify something that serious? The answer stopped me cold. He said he just did not like him. That was it. No data. No documentation. No attempt to dress it up as an operational concern. Just personal dislike, spoken plainly, as if that alone should be enough.

What made it even harder to process was who Ahmad was. He was one of the strongest Process Assistants I had ever worked with. His impact on the Driver Trainer program alone set him apart. Ahmad had a rare ability to connect with Delivery Associates, meeting them where they were and walking them through safety expectations, delivery standards, and Amazon's increasingly complex procedures with patience and clarity. He did not rush people. He did not posture. He taught. His work was precise, consistent, and deeply respected by the drivers he supported. You could see the results on the floor. Better compliance. Fewer issues. Higher morale.

Hearing that kind of baseless hostility directed at someone so essential to the operation felt like a slap in the face. Not just to Ahmad, but to the values Amazon claimed to stand for. Customer obsession. Ownership. Bias for action grounded in data. None of that mattered in that moment. Personal bias had overridden merit without hesitation. It was a blunt reminder that beneath the language of metrics and leadership principles, decisions were still being driven by ego and preference. And once you saw it clearly, it was impossible to unsee.

I stared at him, my mind racing. Internally, the thought was immediate and unfiltered: what a fucking asshole. Outwardly, I kept my reaction neutral, even dismissive, the kind of shrug that probably led him to believe I would handle it the way he wanted. There was never a chance of that happening. I was not going to betray Ahmad to satisfy someone else's ego. Instead, I made a quiet decision to get him out of a toxic environment before it swallowed him.

Without saying a word to Ahmad about what had just happened, I reached out to a few leaders I trusted at other stations. People I knew still cared about competence more than politics. I asked around discreetly, framing it as career development, not an escape plan. One of them had an opening, but there was a catch. Ahmad would need to pass a POD. It was a hurdle, but it was also a door.

When I finally brought it to Ahmad, I positioned it exactly that way. An opportunity. A step forward. I never told him about the leader's malicious intent or the conversation that sparked all of this. We focused entirely on preparation. We sat down and worked through POD questions together, broke down how to structure responses, and ran mock interviews until his answers felt natural instead of rehearsed. Ahmad took it seriously. He always did. When the day came, he passed his POD. Even better, what I had assumed would be a lateral move turned out to be a promotion. Watching him advance and leave that station behind was one of the most satisfying moments of my time at Amazon.

That feeling did not last forever. A couple of years later, I heard that Ahmad had been terminated for a minor safety violation by yet another incompetent leader. The news hit hard. Someone who worked as carefully and consistently as Ahmad did not deserve to be discarded so casually. It was devastating, but not surprising. His story became another example of a pattern I had seen too many times. A leadership culture that talks endlessly about standards and principles, yet routinely misuses power and wastes talent.

Ahmad's resilience, even in the face of that system, stayed with me. His success, even if it was cut short, was earned. It was real. And it stood in stark contrast to a leadership culture that too often failed the very people who made the operation work in the first place.

By the time Ahmad's story ended the way it did, any illusions I still carried about fairness inside Amazon were gone. What remained was a clear pattern I could no longer rationalize away. Competence did not protect you. Integrity did not protect you. Even results did not protect you if the wrong person decided you were inconvenient. Talent was not evaluated in isolation. It was filtered through ego, insecurity, and unchecked authority, and once that filter turned against you, the outcome was usually predetermined.

What haunted me most was how predictable it all became. Ed. Ahmad. Kassie. Scott. Different roles, different stations, different circumstances, but the same underlying mechanics. People who questioned unsafe ideas, who cared deeply about doing the job right, or who simply refused to play politics were slowly isolated, undermined, or removed. Not because they failed the operation, but because they exposed the weaknesses of the people above them. The system did not break accidentally. It defended itself by eliminating friction, and friction increasingly meant anyone with a backbone.

I started to understand that Amazon's greatest strength, its relentless drive for efficiency, had quietly become one of its greatest liabilities. When everything is optimized for speed, scale, and control, there is very little tolerance for humanity, dissent, or nuance. Leaders who should have been coached out were protected. High performers who should have been elevated were discarded. And the burden of holding the operation together fell, again and again, on people willing to absorb pressure without protection.

What made it harder was knowing how much better it could have been. I had seen what healthy leadership looked like. I had built teams that trusted one another. I had watched DSPs and associates rise when they were supported instead of threatened. None of that was theoretical. It worked. But it required courage, humility, and accountability, qualities that too often evaporated as people climbed higher.

By the end of my time at Mission Viejo, I wasn't just tired. I was clear. Clear about what Amazon rewarded, what it tolerated, and what it quietly destroyed. Clear about the cost of staying silent versus the cost of speaking up. Clear about the fact that I had spent years standing between a machine and the people it consumed, believing I could soften the impact if I worked hard enough or cared deeply enough.

That belief finally broke. Not because I stopped caring, but because I realized the system was never designed to be softened. It was designed to extract. And the longer you stayed, the more it asked you to compromise pieces of yourself you could never fully get back.

That realization didn't come all at once. It settled slowly, built from moments like these, from names and faces and careers cut short for reasons that had nothing to do with performance. And once it settled, it changed how I saw everything that came next.

Fixing What the So-Called Expert Broke

Dispatch operations, what most of us still called Load Out, continued to evolve at the Rancho Sereno station as we worked to improve both efficiency and safety. Load Out Associates used twenty-watt blowhorn sirens to signal when vans were entering and exiting the drive lanes, a necessary safety measure meant to keep associates from accidentally stepping into active traffic. It helped reduce near misses, but it did nothing to solve the performance problem. On Time Dispatch metric sat stubbornly around 50%, meaning nearly half the vans were leaving late and putting customers' Promised Delivery Dates at risk.

After spending time observing the floor and talking directly with Delivery Associates, the real bottleneck became clear. DAs were losing valuable minutes walking back and forth to retrieve racks staged far from their parked vans. Those lost minutes added up quickly and were made worse by the All In All Out safety policy. No van could leave until every van was fully loaded and every DA was seated. The policy existed for good reason. It had been put in place after serious near miss incidents at other stations. But in practice, it turned the slowest load into the pace setter for the entire wave, holding dozens of vans hostage to a single delay.

Rather than pushing drivers harder, I looked for unused capacity elsewhere. During the final thirty minutes of the night sort shift, many Sort Associates had little productive work left. I reorganized them and positioned them along the North and South drive lanes to help move racks and load vans, allowing DAs to stay focused on their routes instead of hauling equipment across the building. The impact was immediate. On Time Dispatch jumped from 50% to 98%.

With results that clear, I pushed leadership to formalize the approach, and it became a standard practice that remains in place today. We introduced Voluntary Extra Time for Sort Associates willing to extend their shifts for additional pay and continue supporting Load Out. Eventually, leadership institutionalized the practice by creating dedicated roles to sustain the improvement. It was another reminder that most operational problems are not solved with pressure or slogans. They are solved by paying attention to where time is actually lost and being willing to rethink how labor is deployed when it matters most.

While the initiative dramatically improved station metrics and stabilized daily operations, it came with a familiar frustration. A senior leader leveraged the process improvement to secure a promotion, while my own contribution went largely unrecognized. It was not an isolated incident. It reinforced a pattern I had experienced before at Amazon, where ideas I built and executed delivered measurable results, but the credit and career advancement flowed upward to those best positioned to claim them. Despite that, I stayed focused on the work itself. Driving real improvement on the floor still mattered more to me than chasing recognition, even as the imbalance became harder to ignore.

The station was periodically visited by the Amazon Customer Excellence Systems team, better known as ACES. On paper, ACES existed to refine last-mile delivery operations by standardizing processes, analyzing data, and introducing tools meant to improve efficiency and customer outcomes. In practice, their impact was uneven. Some ACES members brought thoughtful, experience-backed ideas that genuinely improved how the station ran. Others arrived with templated solutions that looked good in a deck but ignored the realities of the floor.

My interactions with ACES were a constant mix of progress and disruption. When their recommendations were grounded in operational context, they could be transformative. When they were not, those same interventions created friction, slowed execution, and forced local teams to spend time undoing or adapting changes that did not fit the station's layout or volume profile. It was another example of a recurring tension inside Amazon, where centralized improvement efforts often collided with on-the-ground reality, and success depended less on the framework being pushed and more on who was doing the pushing and how well they understood the operation.

After one particular ACES visit, the station's On Time Dispatch numbers began behaving erratically. OTD would dip without a clear operational reason, leaving leadership confused and looking for someone to blame. As usual, the Load Out team became the immediate scapegoat. Rather than argue from instinct, I built a data model that tracked Delivery Associate exit times against the station clock. The results were clear. Vans were leaving when they were supposed to. There was no execution gap on the floor.

Yet metics dashboard continued to report poor OTD performance. That disconnect pushed me to dig deeper. Eventually, I found the real culprit. During the ACES visit, adjustments had been made to the station's WiFi and Bluetooth beacon configuration. Those beacons are critical. They ping the DAs' Rabbit devices and generate the timestamps used to calculate entry and exit times. By changing beacon placement and signal behavior without fully understanding the downstream impact, ACES had unintentionally distorted the timing data. The system thought vans were late when they were not.

The experience was a perfect example of how well-intentioned changes can quietly undermine an operation. ACES exists to drive improvement, and in many cases they do. But this incident reinforced a hard lesson. Even small technical adjustments, when not fully vetted against existing workflows and systems, can create false failures that erode trust and misdirect accountability. Innovation only helps when it is grounded in operational reality. Otherwise, it just creates noise that the people on the floor are left to absorb.

To calculate On Time Dispatch, Amazon relies on three different methods to capture when a driver actually leaves the station. Each method varies in accuracy, and understanding the differences explains why OTD can look solid one day and broken the next.

The most accurate method uses Bluetooth beacons. These are small, battery powered devices placed strategically throughout the station and yard. Each beacon emits an ID signal every second within roughly a fifty-yard radius. When a driver logs into their Rabbit device with Bluetooth enabled, the device begins registering those signals. The OTD clock starts with the first beacon ping recorded in the loading area, which marks the driver's arrival, and stops with the final ping as the driver exits the station. Because the system tracks movement through multiple physical points, this method provides the most precise measurement of how long a driver actually spent in Load Out.

When the beacon system is unavailable or compromised, the Rabbit geo fence becomes the fallback. This method relies on a virtual perimeter set approximately eight hundred twenty feet from the station. When a driver's Rabbit crosses that boundary, the system records the dispatch time. While serviceable, this approach includes a built-in five-minute buffer, which introduces small but meaningful discrepancies between what the system reports and what actually happens on the ground.

The third and least reliable option is known as the Calculated method. This is used when neither beacon data nor Rabbit geo fence data is available. In this case, dispatch time is inferred by working backward from the first delivery stop. The system subtracts the planned stem time, which is the travel time from the station to the first stop, along with an additional fifteen minute fueling buffer. While this keeps the metric alive, it is highly susceptible to distortion. A driver delayed before their first stop can appear on time, while another who leaves cleanly but hits traffic can appear late.

Bluetooth beacons are the preferred standard because they capture four critical timestamps. Arrival on station property, entry into the loading zone, exit from the loading zone, and departure from station property. Those data points give the clearest picture of what actually happened during dispatch. The fallback methods exist to maintain continuity, but they trade precision for coverage. When leadership does not understand which method is driving the number, OTD stops being a performance metric and becomes a guessing game.

At our station, an ACES team member had pushed the beacon configuration far beyond what the system was designed to handle. The signal radius was cranked up to its maximum setting, and additional beacons were mounted on light poles throughout the parking lot and along the exterior of the building. The result was absurd. Vans were still being logged as inside the station long after they had physically exited. In one case, a beacon continued pinging a van that had stopped at a gas station nearly a quarter mile away. With the system configured that way, hitting OTD was mathematically impossible. The metric was broken before the first van ever left.

The damage went beyond bad data. Leadership reacted to the numbers, not the cause, and the station was immediately put on the defensive. What looked like poor execution was actually a technical misconfiguration introduced by a team that had already moved on. It was a textbook example of how well-intentioned changes, when made without understanding downstream impact, can quietly sabotage an operation and leave local teams chasing ghosts.

That incident reinforced a troubling pattern I had begun to recognize. Decisions made by teams outside the station were directly undermining the very success they were supposed to support. It was frustrating on a deeper level because it ran counter to a principle that had been drilled into me early in my career, especially during my time in Portland. You leave things better than you found them. You set the next team up to succeed, not to explain away failures they did not create. Watching that standard erode, replaced by disconnected experimentation and unowned consequences, was both disheartening and telling.

What made this shift even harder to ignore was the timing. These issues began surfacing not long after Jeff Bezos stepped down as CEO and Andy Jassy took over. Under Jassy's leadership, the culture inside Amazon felt different on the ground. Decisions that once seemed tightly connected to operational reality started to feel more detached, more theoretical, and more willing to accept collateral damage at the station level.

The alignment and foresight that had previously anchored Amazon's execution began to erode. Teams made changes without fully owning the downstream impact. Metrics were trusted more than context. Local operators were left explaining failures they did not create. Taken together, it raised a troubling question. As Amazon continued to scale in an increasingly complex environment, was it losing the operational discipline and judgment that once set it apart, or was this simply the cost of leadership drifting further away from the work itself.

When the ACES team member returned the following week, I pulled him aside to talk through the changes he had made and the damage they had caused. I was looking for solutions, or at the very least, accountability. What I got instead was one of the most unprofessional interactions of my career. For fifteen straight minutes, he went on a rant, dismissing my concerns and telling me I was "too old" to understand the new systems. He followed that up by telling me to mind my own business, peppering the conversation with foul language and pure arrogance. It was insulting, condescending, and completely out of line. Holding my tongue took every bit of restraint I had, and even then, the encounter stayed with me long after it ended.

That moment went far beyond a personal clash. It crystallized a deeper cultural shift I was already starting to see inside Amazon. This was a company that once took pride in debate, mutual respect, and working backward from the realities on the floor. What I experienced instead was ego masquerading as expertise and contempt directed at anyone who challenged it. The gap between field operations and external teams had grown wide enough that basic professionalism was no longer guaranteed.

Looking back, that interaction stands out as a turning point. It symbolized a workplace dynamic that was changing for the worse, where collaboration and shared ownership were being replaced by arrogance and detachment. It forced me to confront a hard question I had been circling for some time. If this was becoming acceptable behavior, what did that say about the culture Amazon was evolving into. That encounter left me deeply frustrated and, more than anything, disappointed. It marked the moment when I stopped believing that everyone involved was still pulling in the same direction. And frankly, God help that punk if I ever see him again!

That was the moment I decided to stop waiting for permission and fix the problem myself. Leaning on my technical background, I accessed the beacon application environment and effectively jailbroke the system to see exactly how it had been configured. From there, it did not take long to understand what needed to change. When I explained what I planned to do to my manager, he looked completely lost, like a confused puppy trying to follow a magic trick. It was obvious he did not fully understand the mechanics of what I was describing, but to his credit, he trusted me enough to give me the green light.

I moved fast. I reconfigured every internal beacon so it only pinged when vans were actually present inside the station and shut down the external beacons that had been mounted on light poles and along the building exterior. The effect was immediate. On Time Dispatch numbers snapped back into alignment with reality, and the chaos created by the earlier misconfiguration disappeared almost overnight. For the first time since the ACES changes, the data matched what was actually happening on the floor.

Once leadership saw Load Out consistently hitting OTD, the pressure eased and the noise finally stopped. The team was no longer forced to defend itself against bad data, and the operation could move forward without chasing phantom failures. It reinforced a lesson I had learned over and over at Amazon. Sometimes the only way to restore order is to step outside the role you are given, take full ownership of the problem, and fix what others either cannot or will not.

The funny part came next. Word got out about what I had done, and within weeks I started hearing from stations across the country. A dozen, at least, reached out asking how to make the same beacon adjustments. What had started as a quiet, unofficial fix at Rancho Sereno turned into an underground support line for other operators dealing with the exact same issue. It was equal parts absurd and telling. Instead of a formal solution being rolled out, the fix spread person to person, operator to operator, the way real knowledge often moves inside Amazon when the system itself refuses to acknowledge the problem.

What struck me most about that moment was how familiar the pattern felt. A problem created far from the floor. Local teams blamed for the fallout. Pressure applied through metrics instead of understanding. And then, quietly, operators fixing things in the margins so the machine could keep moving. The official narrative never changed, but the reality on the ground did, because people who actually understood the work refused to let bad decisions sink the operation.

That underground spread of the beacon fix said more about Amazon than any leadership principle ever could. Knowledge did not flow through formal channels because formal channels were too slow, too political, or too invested in being right. It flowed sideways, from station to station, manager to manager, carried by people who were tired of watching preventable failures get pinned on teams that were already stretched thin. That was how survival worked inside Amazon by then. You learned who actually knew what they were doing, and you listened to them, regardless of title.

By that point, I was no longer surprised when real solutions lived in backchannels instead of playbooks. The organization still talked about standardization and best practices, but in practice, those only worked when they were shaped by people close enough to the work to feel the consequences. Anything imposed without that understanding became noise at best and sabotage at worst.

Fixing the beacon system should have been a straightforward operational correction. Instead, it became another example of how far Amazon had drifted from its own roots. Debate was no longer welcomed if it threatened ego. Experience was discounted if it did not come wrapped in a new framework. And respect, once a baseline expectation, had become optional, especially when power was uneven.

That realization stayed with me. I had spent years believing that if you showed results, protected people, and solved real problems, the system would eventually recognize it. What Rancho Sereno finally made clear was that recognition was never the point. The system did not reward truth. It rewarded alignment. And alignment increasingly meant staying quiet, accepting bad decisions, and letting others take credit for cleaning up the mess.

I kept doing the work because the people on the floor deserved someone who would. The drivers, the associates, the DSP owners, they were the ones absorbing every mistake made above them. But internally, something had shifted for good. I stopped believing that fixing one more problem, delivering one more save, or holding one more line would change the direction Amazon was heading.

The machine would keep running. It always did. The question I could no longer ignore was whether I wanted to keep being the one standing between it and the people it chewed up along the way.

Persistent Operational Improvements

At that point, Amazon had turned its attention sharply toward tightening the quality of DSP delivery execution, and the timing could not have been better for me to step in and make a real difference. I was coming into this phase with firsthand exposure to what Delivery Associates dealt with every single day, not in theory but on the floor, in the yard, and out on the road. I had seen how small breakdowns multiplied into missed deliveries, safety shortcuts, and unresolved customer complaints. What stood out most was that many of these issues were not caused by a lack of effort or care, but by the absence of clear, enforceable standards that everyone could actually follow. The system relied too heavily on tribal knowledge and local workarounds. I saw a clear opening to bring structure where there was ambiguity, and consistency where there was constant variation, by building comprehensive Standard Operating Procedures across the areas that were hurting the operation the most.

One of the first problems I went after was Product Not on Van, or PNOV, those moments when a package was accidentally left behind during load out and only discovered after the driver was already on route. PNOVs were deceptively disruptive. A single missed package could trigger failed deliveries, customer complaints, reattempts, and extra work for already stretched teams. I built an SOP that broke the process down into clear, repeatable steps, from how PNOVs were identified, to how they were logged, escalated, and resolved as quickly as possible. The goal was speed and accuracy without chaos. By standardizing the response, we reduced delays and eliminated the scramble that usually followed these discoveries, turning what had been a daily frustration into a manageable exception.

Safety was the next area that demanded attention, especially given Amazon's zero tolerance posture when it came to risk. Rather than treating safety as a checklist exercise, I created detailed Safety SOPs that reinforced expectations at every stage of the day. That included pre departure vehicle inspections, proper load securement, and on road behaviors that reduced preventable incidents. What mattered to me was not just compliance, but understanding. I wanted drivers to know why these steps existed, how they protected them, and how cutting corners almost always came back to hurt someone later. The intent was to move safety from something drivers felt policed for, into something they took ownership of because it made sense in the context of their day.

Performance improvement was another area where structure made a tangible difference. I developed SOPs that helped stations and DSPs track delivery efficiency in a way that was practical, not abstract. These processes gave managers a consistent framework for coaching, grounded in real data rather than gut instinct or one off observations. Instead of vague feedback, drivers received specific guidance tied to measurable behaviors. That shift changed the tone of performance conversations. It made improvement feel achievable instead of punitive, and it gave DSPs tools they could actually use to drive consistency across their teams.

Delivered Not Received cases were another chronic pain point in last mile operations. From the driver's perspective, nothing was more demoralizing than being blamed for a missing package they knew they had delivered. From the customer's side, the frustration was just as real. I introduced a standardized approach to investigating DNRs that balanced accountability with fairness. The process focused on facts, delivery data, photos, and patterns rather than assumptions. This not only improved resolution times for customers, but also protected drivers from being unfairly penalized when the evidence did not support the claim.

One of the most meaningful changes came through my work on DSP Scorecards. These scorecards already existed, but too often they functioned as static report cards rather than tools for improvement. I reworked them to include more relevant and actionable indicators, such as on time delivery performance, package accuracy trends, and customer experience signals. The intent was to move beyond surface level metrics and give DSPs visibility into what was actually driving their results. When the scorecards became diagnostic instead of punitive, they started prompting real operational conversations and data backed decisions instead of defensiveness.

Taken together, these SOPs created a foundation for more consistent, higher quality execution across the station. More importantly, they gave both managers and drivers clarity in an environment where expectations were relentless and margins for error were thin. Standardization was not about control for its own sake. It was about removing guesswork, reducing unnecessary friction, and giving people the structure, they needed to succeed in a system that rarely slowed down for anyone. In a network built on speed and precision, these processes became the quiet backbone of operational stability and sustained performance.

What mattered most to me in that phase was that the work finally felt aligned. The SOPs were not theoretical exercises built to satisfy audits or check boxes. They were grounded in what I had watched fail in real time and what I knew drivers, DSPs, and station teams were actually capable of executing under pressure. Every document was written with the assumption that the reader was exhausted, overloaded, and operating inside a system that rarely forgave mistakes. Clarity was not a nice to have. It was survival.

There was a quiet shift that followed. Fewer panicked escalations. Cleaner handoffs. More consistent behavior across DSPs that had previously operated in completely different ways under the same roof. Drivers stopped guessing. Managers stopped improvising. When something went wrong, and something always did, there was a shared language for how to respond instead of ten competing interpretations of policy. That alone reduced friction more than any motivational speech or metric ever could.

Ironically, this was also the moment when I felt the gap between contribution and recognition widen the most. The SOPs worked. Quality improved. Noise dropped. But the system did what it always did. It absorbed the improvement and kept moving without slowing down to ask who built the scaffolding holding it together. By then, I had stopped expecting it to. The satisfaction came from knowing that fewer drivers were being blindsided, fewer DSPs were being whiplashed by inconsistent expectations, and fewer customers were caught in preventable failures.

Looking back, that body of work represents the clearest expression of how I believed Amazon should operate. High standards paired with clarity. Accountability paired with fairness. Structure that supports people instead of trapping them. The tragedy is not that those ideas failed. It is that they were treated as optional rather than foundational.

By the time those SOPs were fully embedded, I understood something I had been circling for years. I was no longer trying to help Amazon become better. I was trying to protect people from the consequences of a system that no longer seemed interested in protecting itself from its own excesses. That realization did not arrive all at once, but it settled in quietly, like a truth you cannot unlearn once you see it clearly.

The work mattered. The impact was real. But the cost of carrying it alone was becoming impossible to ignore.

Pillars of Continuous Improvement

Amazon's commitment to continuous improvement is not a slogan or a side initiative. It is the backbone of how the company actually operates, day in and day out. This mindset influences everything, from how logistics networks are designed to how leaders think about performance and accountability. Efficiency, innovation, and customer satisfaction are not treated as competing priorities but as interconnected outcomes of the same system. Unlike traditional corporate policies that get written once and rarely revisited, Amazon's approach is built on a living framework. Processes are expected to evolve as conditions change, as technology advances, and as real operational pain points surface. The goal is not simply to keep the machine running, but to keep refining it, tightening it, and pushing it further than competitors are willing or able to go.

What makes this framework durable is that it is grounded in adaptation, not perfection. Amazon assumes that today's best process will eventually become tomorrow's bottleneck. That assumption drives a constant cycle of inspection, adjustment, and reinvention. Teams are encouraged, and often pressured, to challenge existing methods, measure outcomes obsessively, and replace what no longer works. This creates an environment where change is not disruptive by default but expected. It also means that operational improvements are rarely final. They are provisional, designed to hold until data, scale, or customer behavior demands the next iteration.

At the center of all of this sits customer obsession, the principle that quietly governs every meaningful decision. Amazon's business model is built around anticipating customer needs and removing friction wherever it exists, even when doing so creates discomfort internally. Innovations like one day and same day delivery were not vanity projects or marketing stunts. They were direct responses to customer expectations that kept rising as Amazon trained people to expect more. Speed, reliability, and convenience became table stakes, and the operation had to bend to support them. Every shift in process, whether it was a small efficiency adjustment or a network wide transformation, ultimately traced back to a simple question. How does this make the customer experience better?

That question acts as both a compass and a filter. Ideas that do not clearly benefit the customer struggle to survive long term, no matter how elegant they look on paper. At the same time, ideas that do serve the customer are relentlessly pursued, even when they strain teams, expose weaknesses, or force uncomfortable tradeoffs. This is how Amazon stays ahead of the curve. It is not by reacting faster than everyone else, but by designing systems that constantly pressure themselves to improve. In that environment, continuous improvement is not optional, and standing still is the fastest way to fall behind.

Standardization is what allows Amazon's operation to function at scale without collapsing under its own complexity. Across a network this large, consistency is not a nice to have, it is the only way to keep execution from fragmenting. At a surface level, Standard Operating Procedures (SOP) and Standards of Work (SOW) can look rigid, even restrictive. In practice, they provide the common foundation that keeps thousands of moving parts aligned while still leaving room for adaptation where it actually matters. By reducing unnecessary variability, these standards create reliability, ensuring that a package delivered in a dense urban core is handled with the same discipline and expectations as one heading down a rural backroad.

What makes this system work is that Amazon does not treat SOPs and SOWs as fixed rules etched in stone. They are living documents, expected to change as the operation learns. Real time feedback from the floor, performance data, and frontline experience all feed into their evolution. When a process breaks or creates unintended friction, it is supposed to be challenged and refined rather than defended out of habit. That constant loop of execution, measurement, and adjustment is what keeps standardization from becoming stagnation. Instead of limiting performance, it becomes the structure that allows teams to move faster, make fewer mistakes, and maintain high standards even as the operation continues to grow and change.

Amazon's obsession with data is what allows the operation to catch problems while they are still small enough to control. Nearly everything is measured, from On Time Delivery rates to defect types, missorts, and customer escalations. That constant flow of data is not collected for reporting theater. It exists so leaders and frontline managers can see trouble forming and intervene before it cascades into missed service promises or burned out teams. With real time analytics always available, adjustments can be made while the day is still unfolding. Bottlenecks get addressed upstream instead of being explained away after the fact. Over time, this turns continuous improvement into a daily discipline rather than a quarterly review exercise.

This mindset aligns closely with the principle of Kaizen, incremental improvement driven by repetition and attention to detail. At Amazon, improvement is not reserved for leadership decks or annual planning cycles. It is expected to come from the people closest to the work. Drivers, process assistants, area managers, and station leaders are all encouraged to identify friction points and suggest fixes. Many of the most impactful changes start small. A tweak to how vans are loaded. A slight adjustment to package handling. A better way to stage routes during peak volume. Individually these changes can seem minor, but over time they compound into meaningful operational gains. That compounding effect is what gives the system its momentum.

What makes this possible is the emphasis on ownership and accountability. These are not abstract leadership slogans inside Amazon. They show up in expectations around how people behave when something breaks. Employees are expected to step in, take initiative, and solve problems as if the operation were their own. The Leadership Principles reinforce this autonomy by pushing decision making downward instead of trapping it behind layers of approval. When something needs fixing, the expectation is action, not escalation for the sake of cover. That pressure can be intense, but it also creates speed. Teams are not just executing tasks. They are actively shaping how the operation evolves, one improvement at a time.

Together, data, Kaizen, and ownership form a feedback loop that keeps the system moving forward. Data exposes the gaps. Kaizen encourages constant experimentation. Ownership ensures someone actually does something about it. In a network as large and unforgiving as Amazon's, that combination is not optional. It is the only way continuous improvement stays real instead of becoming just another corporate phrase.

Technology and automation have become some of the most powerful accelerators behind Amazon's push for continuous improvement, but they are not deployed for novelty or optics. Robotics inside fulfillment centers, AI driven inventory systems, and machine learning models that optimize delivery routing all serve a very specific purpose. They remove friction from manual processes while allowing the operation to scale far beyond what human coordination alone could support. Speed is part of the equation, but it is not the only goal. These systems are designed to improve accuracy, reduce errors, and enhance safety, all while absorbing relentless growth in volume. The real advantage is not that the work moves faster, but that it moves more consistently, with fewer breakdowns as demand spikes.

What keeps this technological engine aligned with reality is the constant presence of feedback loops. Amazon pulls insight from nearly everywhere it can. Customers, employees, DSP owners, delivery associates, vendors, competitors, news coverage, and even social media all feed into how processes are evaluated and adjusted. Drivers surface real world constraints that routing algorithms do not always see on paper. Customer feedback exposes where service expectations are being missed.

DSPs highlight where policy decisions create unintended operational strain. None of this input exists in isolation. It gets aggregated, analyzed, and folded back into process changes that ripple across the network. This is how the system stays responsive instead of rigid, and why improvements rarely come from a single source or perspective.

Continuous improvement at Amazon is not an abstract concept discussed in leadership meetings. It shows up in the physical design and daily rhythm of the operation. Delivery stations operate on a nonstop cycle engineered to maximize throughput while minimizing congestion. Picking, staging, and dispatching are sequenced to keep volume flowing instead of stacking up. DSP load out times are staggered to prevent yard gridlock and reduce stress during peak moments. Warehouse layouts are deliberately structured with wide central aisles, strategically placed docks, and cross docking zones that shorten travel paths and reduce unnecessary handling. Even changes that seem minor, like adjusting parking zones or reconfiguring traffic flow, can shave minutes off each route and reduce compounding delays over the course of a day.

Taken together, these refinements illustrate how deeply improvement is embedded into the operation. Nothing is too small to examine, and nothing is assumed to be finished. Technology provides the leverage, feedback provides the direction, and daily execution provides the proof. In an environment where volume never stops and expectations only rise, continuous improvement is not something Amazon schedules time for. It is something the operation is built to do by default.

Beyond the mechanics of logistics and process design, Amazon also understands that continuous improvement breaks down quickly if the people doing the work feel disposable or disconnected. Team engagement is treated as another operational input, not a soft afterthought. Small initiatives like Muffin Mondays, hydration drives, or simple recognition events may look insignificant from the outside, but inside high pressure delivery stations they matter. They signal that leadership sees the strain of the work and is at least attempting to acknowledge it. In environments where the pace is unforgiving and turnover is always a risk, those moments of normalcy and appreciation can make a long shift feel a little more human.

At a broader level, Amazon also makes a point of engaging with the communities that surround its facilities. Participation in beach cleanups, food drives, and local volunteer efforts is part of how the company frames its role beyond being just an employer or a logistics machine. These initiatives are not purely symbolic. They help ground massive operations in real places with real people, and they give employees a sense that their work connects to something larger than daily metrics and dashboards.

Taken together, these efforts contribute to a culture that recognizes improvement is not driven by systems alone. Morale, connection, and shared purpose all influence how well processes actually hold up under pressure. While these initiatives do not erase the intensity of the work, they do create small points of stability in an otherwise demanding environment. In a system built on constant motion and measurement, those human touches help sustain the people who ultimately keep the operation moving.

Safety is another core pillar of Amazon's operational philosophy, and it is deliberately embedded into daily routines rather than treated as a box to check for compliance. The expectation is that safety lives on the floor, not in a policy document. Safety Circles create space for associates to speak openly about risks they see in real time and to propose practical improvements based on lived experience. Weekly audits and standing Safety Committees add another layer of accountability, ensuring that concerns are not only heard but tracked, revisited, and addressed. Tools like the "Why I Stay Safe" boards personalize what can otherwise feel abstract, turning safety from a list of rules into something tied directly to people's lives, families, and long term well being. Role specific training reinforces this mindset by giving employees the knowledge and confidence to spot hazards early and intervene before small issues become serious incidents. In this environment, safety is not reactive. It is proactive, visible, and culturally reinforced.

Strong managerial oversight is what binds all of these elements together and prevents continuous improvement from drifting into theory. Managers are expected to live inside the data, tracking key performance indicators like Delivery Completion Rate and On Time Dispatch to understand where execution is slipping and why. These metrics are not viewed in isolation. They are used as signals that prompt deeper conversations about process gaps, staffing challenges, training effectiveness, or safety risks. Weekly meetings with DSPs reinforce this discipline, creating a regular forum to review performance, surface concerns, and align on corrective actions. The emphasis on transparency and accountability keeps improvement efforts grounded in reality rather than optimism. When something is not working, it is visible, discussed, and owned.

Taken together, these pillars of continuous improvement form the operational spine of Amazon's success. They allow the company to deliver at massive scale while maintaining a level of consistency that many competitors struggle to achieve. More importantly, they create a system that can adapt as conditions change, whether that change comes from rising customer expectations, shifting labor dynamics, or new technological constraints. By embedding these principles into every layer of the operation, Amazon does more than maintain efficiency. It continuously reshapes what operational excellence looks like, often forcing the rest of the industry to chase standards it has already moved past.

Where this system begins to strain is not in its design, but in how unevenly it is applied. Continuous improvement, when practiced with discipline and humility, is powerful. When practiced selectively or weaponized through metrics, it becomes corrosive. The same mechanisms that surface defects and inefficiencies can just as easily be used to deflect accountability, justify bad decisions, or pressure teams to absorb strain without addressing root causes. Improvement loses its integrity when data is valued more than context and when speed is rewarded at the expense of judgment.

I saw this tension play out repeatedly. The framework itself was sound. The tools worked. The principles made sense. But the outcomes depended almost entirely on who was interpreting them and why. In the hands of leaders grounded in operations, continuous improvement led to smarter processes, safer execution, and stronger teams. In the hands of leaders chasing optics, promotions, or control, the same framework became punitive, detached, and quietly destructive. The difference was never the system. It was the character and competence of the people wielding it.

That distinction matters because Amazon's scale magnifies everything. Good decisions ripple outward and compound into real advantage. Bad decisions do the same, just in the opposite direction. When leadership loses touch with the floor, continuous improvement stops being a shared discipline and starts feeling like a moving target imposed from above. Standards change without explanation. Metrics shift without warning. Accountability flows downward while ownership evaporates upward. Over time, that disconnect erodes trust, and once trust is gone, even the best-designed systems struggle to hold.

What I ultimately came to understand is that Amazon's greatest strength and its greatest vulnerability are the same thing. The company is built to move fast, to iterate relentlessly, and to push limits without waiting for permission. That engine can produce extraordinary results, but it also requires leaders who know when to slow down, listen, and protect the people inside it. Continuous improvement is not just about fixing processes. It is about knowing which problems are worth fixing, which pressures are necessary, and which ones are quietly breaking the system from the inside.

For all its sophistication, Amazon still runs on people. Drivers making judgment calls at front doors. Associates moving volume through narrow windows. DSP owners betting their livelihoods on promises that shift beneath their feet. Managers holding the line between what the system demands and what humans can realistically sustain. When continuous improvement honors that reality, it works. When it ignores it, the cost shows up later, often in ways no dashboard is built to capture.

Looking back, I do not question the philosophy. I question the discipline with which it was upheld. Continuous improvement is only as ethical as the leaders enforcing it. Without empathy, context, and accountability at the top, it stops being a path toward excellence and becomes a justification for excess. Amazon proved what is possible when systems are designed to learn. It also revealed how fragile those systems become when learning gives way to ego, distance, and unchecked pressure.

That tension sits at the heart of everything I experienced. The framework was never the problem. The question was always whether leadership was improving the system, or simply demanding that people adapt until something finally broke.

The Only One Who Believed

Arnold was one of my Process Assistants who stood out immediately, not because he talked the loudest or played politics well, but because he consistently carried the operation when things were falling apart. He had a deep command of the Dispatch process and understood it end to end, including all the ugly edge cases that never make it into standard work. When volume spiked, routes broke, drivers called out, or systems lagged, Arnold did not panic or deflect. He absorbed the pressure and kept the flow moving. He thought clearly in chaos, adapted on the fly, and solved problems in real time without needing hand holding or validation. His communication was precise, rooted in facts, and focused on execution, not excuses.

What many people never saw was how much weight he carried quietly. Arnold had come up the hard way, shaped by a life that demanded resilience long before Amazon ever did. That background showed up in how he worked. He did not waste energy posturing or seeking attention. He showed up prepared, stayed composed, and handled responsibility with a seriousness that came from knowing what instability actually looks like. In an environment defined by unpredictability, Arnold was steady. Knowing he was on shift brought a level of confidence and calm that only comes from working with someone who has been tested long before the moment demands it.

What impressed me most was his composure. Amazon's last mile environment grinds people down. It rewards urgency while punishing mistakes, and it rarely offers relief. Arnold never let that grind turn into bitterness or disengagement. He did not posture. He did not complain. He took ownership, solved problems, and moved on to the next fire. That kind of resilience is not teachable, and it is incredibly rare. Over time, it became obvious that Arnold was already operating above his role. He was making judgment calls that impacted downstream performance, coaching others in the moment, and quietly filling gaps leadership either could not see or chose not to address. To me, promoting him was not a gamble. It was an overdue correction.

I also understood the reality behind the title. Amazon chronically underpays its PAs relative to the responsibility they carry, especially in last mile operations where mistakes have immediate customer impact. Arnold was supporting his family on compensation that did not come close to reflecting his contribution. A promotion was not just about career progression. It was about fairness, stability, and acknowledging that sustained excellence deserves more than verbal praise and empty thank you. I took that responsibility seriously. I put together a detailed case for leadership that laid out Ronald's performance, his problem solving, his operational impact, and his readiness to operate at the next level. There was nothing vague or emotional about it. It was data, observation, and firsthand operational evidence.

Leadership dismissed it almost instantly. There was no real discussion, no curiosity, no attempt to understand the operation from the ground up. The decision felt predetermined, as if recognizing real talent was an inconvenience rather than a priority. This was not an isolated failure. It was a pattern I had seen repeatedly at Amazon. Promotions were often influenced less by operational excellence and more by visibility, timing, and whether someone fit leadership's narrow idea of what growth was supposed to look like. Many managers making these decisions had limited hands on experience in last mile operations. They spoke fluently in metrics and narratives but lacked the situational awareness to recognize who actually carried the business day to day. As a result, people like Arnold were overlooked while less capable but more politically aligned individuals moved ahead.

That refusal to recognize talent is one of Amazon's quiet dysfunctions. The company talks endlessly about ownership and raising the bar, yet routinely ignores the people who embody those principles most consistently. It burns out its strongest operators, underpays them, and then acts surprised when they disengage or leave. In Ronald's case, leadership's failure was not just short sighted. It was negligent.

I was frustrated, but not surprised. I had learned that advocating for the right people inside Amazon often meant hitting an immovable wall. Still, I refused to abandon him. If leadership would not invest in Ronald, I would. I continued to support him, coach him, and make sure his contributions were visible wherever I could. He deserved far better than what the system was willing to give, and his story was yet another reminder that at Amazon, talent does not rise on its own. It has to fight its way up, often despite the very leaders who claim to value it.

Just days later, Ronald's value was no longer theoretical. It was tested in real time, under conditions where there was no playbook and no margin for error. A Serv 1 network failure crippled our internal communication systems just thirty minutes before dispatch. Rabbit devices became useless overnight. Delivery Associates could not see their staging locations, could not scan packages, and could not access GPS routing. The operation was effectively blind. With the clock running down and hundreds of drivers standing idle, the choice in front of me was stark. Delay dispatch and accept a cascade of late deliveries, missed promises, and downstream penalties, or attempt a manual workaround with incomplete information and very little time.

While leadership would have defaulted to escalation emails and conference calls, Arnold stayed focused on the floor. He recalled something critical from earlier that morning. During a deep dive into routing, he had noticed that even though only the first wave had physically been staged, every route had already been logically assigned to a staging area in the system. That detail mattered. It meant the data still existed, even if the interface to access it was down. That single insight changed everything.

I had already downloaded a full routing manifest earlier in the shift, something I did out of habit because experience had taught me systems fail at the worst possible moments. The manifest contained route assignments, package lists, stop order, and customer addresses. With more than four hundred Delivery Associates impacted, we knew we could not recreate the entire day manually. Arnold helped cut through the noise. We focused on the first five stops for each route. That was enough to get drivers moving, clear the yard, and buy time. We split the manifest by DSP, highlighted the priority stops, and pushed the information out through Amazon's real-time communication platform app. DSP leadership then relayed the details to their drivers, who manually entered addresses into their personal phones just to get rolling.

It was not clean. It was not elegant. But it worked. Dispatch went out. Vans cleared. Routes began moving. About an hour later, the technical team restored the Serv 1 system and normal operations resumed. From the outside, it looked like a routine recovery. Inside, it was anything but. Without Ronald's situational awareness and willingness to think beyond standard work, we would have lost the entire shift. Every package still made it to the customer on time. No service promise was broken. No executive escalation followed.

The most telling part came afterward. The workaround Arnold helped design was documented and later adopted as a contingency SOP by other stations. It became institutional knowledge. Yet once again, the person who had made the difference received no meaningful recognition from leadership. No urgency to revisit his promotion. No reassessment of the earlier dismissal. The system absorbed his contribution, standardized it, and moved on.

That moment crystallized everything that was broken about Amazon's approach to talent. Arnold demonstrated exactly the kind of ownership, bias for action, and customer obsession the company claims to reward. Under pressure, he protected the business when leadership systems failed. And still, leadership could not see him. Not because the evidence was missing, but because the culture was never built to recognize people like him in the first place.

Unbeknownst to me at the time, several senior Amazon leaders from Seattle were visiting the station that day. Later that afternoon, one of them pulled me aside to talk about the morning's chaos. He had clearly been briefed, and the conversation quickly moved past surface level pleasantries into a candid discussion about what actually happened. As we talked, the subject drifted toward Amazon's leadership culture. To my surprise, he did not defend it. Instead, he openly acknowledged its flaws and admitted he was seriously considering leaving the company altogether in search of a role that aligned more closely with his personal values. That alone said more than any leadership principal poster ever could.

Then he dropped the part that stopped me cold. A station leader had already taken credit for resolving the dispatch crisis. Presented it as their save. Their call. Their execution. He followed that up by saying he knew that was bullshit. He knew the solution did not come from them. He knew it came from me. That validation mattered, not because I needed my ego stroked, but because it confirmed what I had lived through time and time again. The people closest to the work rarely get the credit, while those furthest from the floor are rewarded for proximity to the story. I immediately corrected the record and made it clear that Arnold had been instrumental. I explained his proactive routing analysis, his situational awareness, and how his insight was the linchpin that made the workaround possible. Without him, the shift would have cratered.

None of this surprised me. By that point, I had lost count of how many operational wins had been quietly stripped from my name and handed to someone else with a better title or a louder voice. It became painfully obvious that competence was optional, but optics were everything. And sitting beneath all of it was something Amazon never openly admits but constantly reinforces. My age, my experience, and my time in the trenches were no longer assets. They were treated like liabilities.

Years of hard-earned judgment, pattern recognition, and firsthand operational scars never translated into advancement or better pay. If anything, they seemed to disqualify me. I watched less capable, less tested leaders move ahead simply because they fit the image Amazon wanted to sell about its future. Younger. Shinier. Easier to package. My contributions were absorbed, cleaned up, and reassigned to people who were skilled at narrating success without ever having to carry it. Another worthless, incompetent asshole got an invisible boost, and the system kept humming along, perfectly comfortable pretending this was meritocracy in action.

A few days later, I found out that Arnold had been selected to participate in Amazon's POD process. I was genuinely happy for him, but I would be lying if I said I did not wonder whether that conversation with the Seattle leader had tipped the scale. I had been through enough PODs to know how selective and political the process could be. I had served as an interviewer. I had coached candidates through it. I knew exactly how brutal it was. Three separate interviewers. Forty-five minutes each. No breaks. No grace. Candidates were expected to demonstrate deep operational fluency, flawlessly map their experiences to leadership principles, and stay composed while being challenged, redirected, and occasionally cornered. It was not just an interview. It was an endurance test designed to filter out anyone who could not perform under sustained pressure.

Arnold could handle it. I had no doubt about that. What lingered was the uncomfortable truth that it took exposure to senior leadership for his talent to be acknowledged at all. Not months of carrying the operation. Not saving a dispatch that could have imploded customer trust. Not doing the right thing when systems failed. It took someone from Seattle noticing, quietly validating the truth, and nudging the machine just enough for it to begrudgingly do what it should have done from the start. That realization said everything about how broken Amazon's leadership pipeline really was.

When I told Arnold he had been selected for POD, I expected excitement, maybe even relief. Instead, what I saw caught me off guard. He hesitated. He looked uncomfortable. There was no celebration in his voice, only uncertainty and a quiet, almost apologetic doubt about whether he belonged there at all. It became painfully clear that this was not someone who had spent his life being told he was capable, let alone exceptional. Arnold had learned to survive by keeping his head down, doing the work, and not expecting recognition. Belief, especially from leadership, was not something he was used to receiving.

I did not let that moment pass. I told him directly that he was ready. Not almost ready. Not lucky to be considered. Ready. I explained why, pointing to what I had seen from him on the floor and under pressure when everything was breaking. I told him I would help him every step of the way, and I meant it. What bothered me was realizing no one else had done the same. Not his managers. Not leadership. Not the system that constantly preached growth and development. If Arnold was going to succeed, it would be because someone finally invested the time to believe in him and backed that belief with action.

Over the next several weeks, we stayed late after shifts, long after the building had quieted down. We worked through POD prep the hard way. We broke down his experiences and rebuilt them into clear, structured stories. I pushed him to see his own work the way an interviewer would, translating real operational impact into language Amazon claimed to value. We practiced answering questions cleanly, then practiced again when I interrupted him mid answer, challenged his assumptions, or forced him to defend his decisions. I taught him how to anticipate follow up questions, how to stay grounded when an interviewer tried to rattle him, and how to speak with confidence without slipping into arrogance. Every session chipped away at the doubt that had been sitting on his shoulders for years.

What mattered most was not just the mechanics of interview prep. It was repetition of belief. Every time he questioned whether an example was good enough, I reminded him it had saved the operation. Every time he minimized his role, I corrected him. I refused to let him undersell himself because I knew exactly how that story usually ended at Amazon. Quiet, capable people get passed over while louder, less competent ones move ahead. I was not going to let that happen to him without a fight.

When the day for the POD finally arrived, Arnold walked in prepared. Not rehearsed. Prepared. He knew his material. He knew his impact. He knew he belonged in that room. And he passed. Cleanly. No drama. No excuses. He earned the promotion, along with a transfer to a station closer to home. That change alone altered his daily life in a way Amazon leadership never fully appreciates. Less commute. More time with family. Less strain. Real, tangible improvement that went far beyond a title change.

On his last day at our station, Arnold found me before he left. He thanked me, but not in the polite, professional way people usually do at Amazon. What he said landed heavier than that. He told me "I was the only person who had ever truly believed in him". Not the only one who helped him prep. Not the only one who supported him professionally. The only one who actually believed he was capable of more.

That moment stayed with me because it said everything about the failure of Amazon's leadership culture. Arnold did not lack talent. He did not lack work ethic. He did not lack resilience. What he lacked was a system willing to see him. In the end, his success did not come from leadership programs, principles, or polished narratives. It came from one person deciding that belief was not optional, and that sometimes the most important thing you can give someone is the conviction that they are not invisible.

That moment stripped everything down to what actually matters. In a place like Amazon, where dashboards, metrics, and relentless pace dominate every conversation, it is dangerously easy for people like Arnold to fade into the background. The system is not designed to notice quiet excellence. It rewards visibility, volume, and proximity to the narrative when something goes right. Ronald's story forced me to confront how much talent gets overlooked simply because it does not self-promote or fit neatly into leadership's preferred mold. It reinforced a hard truth. Success is not just about hitting numbers or surviving the day. It is about recognizing the people who are carrying the weight of the operation and making sure they are not left behind.

What changed Ronald's trajectory was not a leadership program or a polished development framework. It was belief, backed by time, attention, and consistency. Once someone finally told him he was capable and treated that belief as real, years of self-doubt began to fall away. His potential did not need to be created. It was already there, waiting for permission to surface. Watching him grow reaffirmed something I have learned the hard way. Leadership is not defined by title, headcount, or visibility. It is defined by whether the people around you are better because you were there. That lesson stayed with me, not because it was dramatic, but because it exposed both the quiet cost of leadership failure and the quiet power of choosing to do the opposite.

What Ronald's story ultimately clarified for me was that Amazon's greatest leadership failures are rarely loud. They do not show up as scandals or public reckonings. They show up quietly, in the form of people who carry the operation without being seen, who solve problems without being credited, and who doubt themselves because no one in authority ever bothered to say, you belong here. That kind of failure does not trigger alarms or escalations. It simply erodes trust, confidence, and continuity one person at a time.

Arnold succeeded not because the system suddenly worked, but because someone interrupted its inertia long enough to insist on a different outcome. That intervention should not have been exceptional. It should have been routine. In a company that claims to value ownership, bias for action, and raising the bar, belief should not be a scarce resource. Yet time and again, I watched belief withheld from the very people who earned it most, while being handed freely to those who mastered optics and proximity.

By the time Arnold left the station, I no longer saw his promotion as a win for Amazon. I saw it as a narrow escape. He had proven what he was capable of despite the system, not because of it. The fact that his trajectory depended on a chain of unlikely moments, a network failure, a visiting leader, and one person willing to fight for him, says everything about how fragile merit really is inside the machine.

That realization stayed with me long after he was gone. It reframed how I understood my own role and limits inside Amazon. I could protect people in moments. I could advocate. I could intervene. But I could not fix a culture that consistently confused visibility with value and silence with weakness. The system did not lack talent. It lacked the will to recognize it without being forced.

Ronald's final words to me were not a victory lap. They were an indictment. If one person believing in him could change everything, then the absence of that belief for so long was not neutral. It was damaging. And once I saw that clearly, I could not unsee it. Leadership, real leadership, was happening in the margins, in late nights, quiet coaching, and unrecorded moments that never made it into performance narratives.

That is where the work actually mattered. Not in dashboards or principles, but in deciding who gets seen, who gets supported, and who gets left behind. Ronald did not need to be fixed. He needed to be recognized. And the fact that recognition came so late, and so narrowly, remains one of the clearest indictments of the system I worked inside.

Leadership Without Lived Experience

The station often felt like a proving ground for experimental process changes, many of which were rolled out with more confidence than competence. New ideas arrived fully branded and confidently announced, yet rarely tested against the reality of how the floor actually functioned. One of the most disruptive examples was the introduction of the so called "All In All Out" Dispatch process.

Under this new system, Delivery Associates were required to enter the station in strict sequence. Vans funneled into a single lane until it filled, then spilled into the next, and continued that pattern until every lane inside the building was occupied. Only after the final van was parked were DAs allowed to exit their vehicles and walk to their staging locations. Departure followed the same rigid logic. One lane exited at a time. No movement was allowed until every driver was in position and cleared.

Leadership framed the change as a safety improvement. On paper, that sounded reasonable. In practice, it ignored the fact that the previous system had already solved for safety without crippling throughput. DAs entered through multiple doors and lanes simultaneously, and no one exited their vehicle until the all-clear siren sounded. It was simple, intuitive, and proven. It kept people safe while allowing the operation to move at the pace required to hit dispatch targets. That balance mattered. And it worked.

The new process did not. Almost immediately, operational efficiency suffered. On Time Dispatch took a hit, and not a small one. Leadership either underestimated or outright dismissed how tightly OTD was tied to fluid movement during load out. Vans stacked up. Time evaporated. What had once been controlled urgency turned into bottlenecks created by design. Over the following weeks, the station clawed its way back toward the revised thirty-minute dispatch target, but only through sheer effort and adaptation by the DAs and frontline leaders forced to make a flawed process usable.

Just as the station began to stabilize under the already flawed "All in All Out" process, leadership managed to make it worse. Another directive came down, this time slashing the dispatch wave time from thirty minutes to an unworkable twenty, without changing a single constraint in the process itself. The decision was clearly made by someone far removed from the floor. DSP owners were furious, and they had every reason to be. Expecting Delivery Associates to load vans and depart in that compressed window without cutting corners or risking injury was detached from reality.

The outcome was exactly what anyone with operational experience could have predicted. Dispatch turned into a frantic race against the clock. Order gave way to chaos. On Time Dispatch, already under strain, dropped even further. Instead of acknowledging the obvious mismatch between expectations and execution, station leadership responded with pressure. More urgency. More demands. More messaging about accountability. What they did not offer was support, flexibility, or a willingness to revisit the decision that created the problem in the first place.

By then, I had stopped expecting help from station leadership altogether. Their pattern was unmistakable. They imposed change, deflected responsibility, and resisted ideas that came from the floor. Whether driven by ignorance, insecurity, or an unwillingness to own bad decisions, the outcome was always the same. The burden of making broken processes work fell downward. I understood that if the operation was going to survive, it would not be because leadership stepped in, but because someone on the ground figured it out anyway.

I started by tearing the dispatch process apart step by step, tracing it from the moment Delivery Associates entered the yard. What became clear was that the new system inside the station was not the primary failure point. Once vans reached the entry lanes, movement was relatively smooth. The real bottleneck existed before dispatch ever began, outside the building, in an area leadership had never bothered to examine.

DAs were being instructed to stage their vans one hundred to two hundred feet away from the north and south entry points. That distance mattered. Each van took roughly one to one and a half minutes just to travel from staging to the door. With more than forty vans feeding in from each side, those minutes stacked up fast. The problem was compounded by the fact that only a single Yard Marshal was assigned to manage traffic flow while also checking Department of Transportation (DOT) paperwork for Step Vans. Predictably, everything slowed to a crawl. To make matters worse, DA arrivals were inconsistent. Some showed up late, others early, and the staggered flow completely undermined any chance of hitting a twenty-minute dispatch window.

Once the root causes were clear, the solution was straightforward. I implemented three changes immediately. First, I moved the staging area to within ten feet of the bay doors and had vans line up, eliminating the wasted travel time entirely. Second, I added a second Yard Marshal to split traffic control and DOT checks, which immediately cut entry delays in half. Third, I worked directly with DSP owners to synchronize DA arrivals. Drivers were instructed to leave their lots ten minutes earlier and arrive as a coordinated group instead of trickling in. That single adjustment removed the variability that had been throwing off the entire sequence.

Getting DSP owners on board was not automatic. They were already exhausted by constant process changes and understandably skeptical of yet another adjustment coming from the station. It took direct, honest conversations to walk them through the data, explain where time was actually being lost, and show how these changes would improve flow without increasing risk or stress on their drivers. Once they saw dispatch stabilize, the hesitation faded. What leadership had failed to do with mandates and pressure was resolved through observation, collaboration, and respect for how the work actually happened.

Once the DSP owners were aligned, the next hurdle was the hardest one. Helping the Delivery Associates adapt. Changing habits in a fast paced, high-pressure environment is never clean, especially when those habits have been shaped by months of inconsistent direction and constant rule changes. The new staging protocol made sense on paper, but in practice it required retraining muscle memory. Coordinated arrivals. Tight staging near the bay doors. Nose to toes instead of spreading out. In the first few days, it was messy. Vans showed up out of sequence. Drivers hesitated. Some reverted to old patterns out of instinct. The early shifts were chaotic as everyone tried to find their footing.

I knew this part could not be fixed with messaging or reminders taped to a wall. It required presence. Every morning, I walked the staging areas myself. I talked directly with drivers, clarified expectations in real time, and corrected issues as they happened instead of waiting for them to show up in metrics later. I worked side by side with the Yard Marshals to keep traffic moving and eliminate confusion before it turned into delays. There was no shortcut here. It took repetition, patience, and consistency. Slowly, the chaos settled. Vans lined up faster. Entries tightened. Small improvements stacked on top of each other until the new process stopped feeling new. What finally made it work was not pressure or enforcement, but giving people the time and support they needed to build the muscle memory to execute instinctively.

By the end of the first week, the difference was undeniable. Vans moved cleanly from staging to their assigned lanes without the drag and hesitation that had plagued dispatch just days earlier. The flow finally matched the design. On Time Dispatch rebounded faster than anyone expected, and for the first time in weeks, dispatch felt controlled instead of frantic. The shift was not just operational. It was emotional. Drivers were visibly relieved. DSP owners felt the pressure ease as predictability returned to their day. What had been chaos started to feel manageable again.

The turnaround was not the result of a single directive or mandate. It came from collaboration, clarity, and persistence. People understood what was expected of them, why the change mattered, and how their role fit into the larger picture. That alignment mattered as much as the physical changes to the yard. Operational fixes only stick when the people executing them are brought into the solution, not treated like obstacles to it.

The numbers told the rest of the story. OTD, which had been sliding steadily, did not just recover. It surged. Dispatch eventually reached a sustained 100 percent, a level of consistency that was rare across Amazon's network. It was proof of what was possible when problems were analyzed honestly and solved with creativity and cooperation instead of pressure. And yet, the response from leadership was exactly what I had come to expect. No recognition. No acknowledgment. No reflection on the fact that the solution came from the floor, not from the decisions that caused the problem in the first place. The operation improved, the metrics looked good, and leadership moved on. Another reminder that at Amazon, results are celebrated, but the people who make them happen are often invisible.

The lack of recognition was demoralizing, but it was also revealing. It reflected a leadership culture at Amazon that consistently failed to recognize real operational achievement because too many of the people placed in decision making roles had little to no firsthand experience actually running a business or managing complex operations. Leadership positions were often filled by individuals fluent in process language and metrics, yet disconnected from the realities of execution. As a result, recognition flowed toward those who complied cleanly and managed optics, not toward those who solved problems when systems broke. When leaders lack operational grounding, they struggle to identify real value, and the work that keeps the business functioning quietly disappears into the background.

Even so, I took pride in what the team accomplished. Faced with a broken directive and no meaningful support, we did not wait for permission that was never coming. We adapted. Through collaboration, persistence, and a relentless focus on execution, we turned a flawed process into something that worked without sacrificing safety or efficiency. That experience reinforced a belief I had learned repeatedly on the floor. Real operational excellence does not come from rigid adherence to orders. It comes from people who understand the work, care about the outcome, and are willing to think beyond the script. It may have been a small victory in the broader machine, but it exposed the widening gap between leadership priorities and the realities faced by those doing the actual work.

That gap stayed with me. It became harder to ignore each time a solution born out of necessity was absorbed by the system without acknowledgment, reflection, or learning at the leadership level. The machine accepted the outcome and discarded the process that made it possible. Over time, that pattern reshaped how I understood Amazon's version of success. Results mattered, but only insofar as they could be claimed without challenging the authority or judgment of those above. When fixes contradicted leadership decisions, silence followed. The work was allowed to stand, but the lesson was never allowed to travel upward.

What made this especially corrosive was how predictable it became. Leadership mistakes created instability. Frontline operators absorbed the damage. Someone on the floor figured it out. Metrics recovered. And then the narrative reset as if the disruption had been unavoidable rather than self-inflicted. There was no postmortem, no recalibration of trust, no reassessment of who should be making decisions in the first place. Each cycle reinforced the same message. Initiative was welcome, as long as it did not expose flawed judgment higher up the chain.

By then, I no longer believed that better data or cleaner execution alone would fix that disconnect. The problem was structural. Too many leaders were insulated from consequence, while too many operators were shielded from recognition. That imbalance did not just distort incentives. It trained people to stop speaking up, stop pushing back, and stop investing emotionally in outcomes they would never be credited for. The quiet cost of that dynamic was enormous. It hollowed out trust and slowly replaced ownership with compliance.

And yet, despite all of that, the work still mattered. The people still mattered. Every time dispatch flowed cleanly, every time a driver finished a route without unnecessary stress, every time a DSP avoided another bruising day of chaos, it reinforced why we kept showing up. Not for leadership approval, but for each other. The real wins happened horizontally, not vertically. They lived in the relationships between operators, drivers, yard marshals, and DSP owners who understood that survival depended on cooperation more than hierarchy.

That chapter at Rancho Sereno cemented something for me. Amazon's systems are powerful, but they are only as good as the people willing to bend them back toward reality when they break. The danger is not that bad processes exist. The danger is when organizations stop listening to the people who fix them. When that happens, excellence does not disappear overnight. It erodes quietly, one ignored solution at a time.

Looking back, that period was not just about dispatch or load out or OTD. It was about learning where real leadership lived and where it did not. Titles did not matter. Approval did not matter. What mattered was whether the work got done without breaking the people doing it. That understanding reshaped how I saw my role inside Amazon and, eventually, why I knew I could not stay forever. When doing right becomes impossible to acknowledge, something has to give.

Built to Be Safe

For the life of me, I have never fully understood why Amazon is so frequently criticized for its safety practices. From my perspective inside the operation, safety, alongside customer obsession, was one of the company's most visible and deeply embedded priorities. From the moment I joined, it was clear that safety was not treated as a slogan or a compliance checkbox. It was woven into daily routines, leadership expectations, and operational decision making at every level. Associates and managers were not simply trained once and left alone. Safety was reinforced constantly. Over the course of my tenure, I became CPR certified more than a dozen times, which alone spoke to how seriously Amazon expected its leaders to be prepared for real emergencies, not just theoretical ones.

Everyone in the warehouse, without exception, received extensive safety training tailored to their specific roles, and those trainings were mandatory. You did not step fully into the work until you passed them. The expectation was not basic compliance but demonstrated competence. That focus did not end once training was complete. Every single shift began with a Start-Up Meeting where safety was front and center. These meetings were not performative. They created space for associates to raise real concerns, many of which led directly to changes on the floor. Processes were adjusted, procedures refined, and new equipment introduced when patterns of risk emerged. Watching that feedback loop function in real time was genuinely impressive and often ignored by critics who never saw how these decisions actually played out inside the building.

That same emphasis extended to Delivery Service Providers and Delivery Associates operating in the last mile. DSPs and DAs underwent rigorous safety training designed around the realities of delivery work, including defensive driving, proper package handling, and situational awareness on the road. For managers, safety did not stop at training completion. We were responsible for reviewing daily safety performance, documenting incidents and near-misses, and looking for patterns rather than assigning blame. These reviews were meant to surface root causes and prevent recurrence. That continuous cycle of review, learning, and adjustment reflected how Amazon approached safety in practice, not just in principle.

Much of the criticism aimed at Amazon's safety practices feels disconnected from the systems I observed firsthand. The framework itself was robust and clearly designed with worker protection in mind. Safety was not an afterthought. It was a foundational pillar of how the operation was run. While no system is perfect and there is always room to improve, the intent and structure I experienced reflected a company actively trying to reduce risk and protect its workforce, even under immense operational pressure.

That commitment was reinforced through the technology embedded in delivery vans. Vehicles were equipped with inward- and outward-facing cameras, GPS tracking, and sensors capable of detecting behaviors such as harsh braking, speeding, seatbelt non-use, rapid acceleration, and distracted driving. When violations occurred, alerts were sent in near real time to DSP Managers and Amazon On-Road Managers, allowing for immediate intervention rather than delayed after-the-fact reviews.

The data generated by those systems became a powerful safety tool. I built daily models to track and analyze safety misses, highlighting patterns like repeated seatbelt violations or excessive speeding. Those reports were shared directly with DSP Owners, and they carried real consequences. Safety performance fed into a DSP's Scorecard, which influenced bonuses, route growth, and overall standing within the program. That direct link between safety behavior and business outcomes created urgency and accountability. Problems were addressed quickly, not buried.

Taken together, the training, technology, and accountability mechanisms created a safety ecosystem far more serious than many outsiders assume. Whatever Amazon's flaws, the systems I worked within showed a consistent effort to reduce risk and protect the people doing the work. That commitment was real, even if it was not always applied intelligently.

This was the contradiction I saw repeatedly. Amazon's safety systems were real, but they were often undermined by leadership decisions made far from the floor. Safety became the justification for top down process changes that ignored how work was actually performed. In the case of "All In All Out," the intent was framed as protection, but the design introduced new pressure points that increased frustration and risk. When rigid sequencing, compressed timelines, and untested workflows collided with real world conditions, the safeguards already in place were weakened, not strengthened. In those moments, safety stopped being a lived practice and became a post hoc explanation, used to defend flawed decisions rather than prevent the problems they created.

One of the most common safety violations I encountered involved Delivery Associates not wearing their seat belts while driving their vans. As I dug into the safety and routing data, a clear pattern emerged. These violations were heavily concentrated in gated communities. The reason became obvious once you looked past the metric. In many of these neighborhoods, homes sit only a few yards apart, creating a constant stop-and-go rhythm. Drivers were in and out of their vans repeatedly, and some began rationalizing that buckling up for such short distances was unnecessary.

Using data from the vans' safety monitoring systems, I was able to pinpoint specific routes and time windows where seat belt compliance consistently dropped. Telemetry, safety alerts, and GPS routing data combined to create a detailed picture of where the problem was occurring. The trends were unmistakable. Certain gated communities showed repeat patterns of non-compliance, allowing me to quantify the issue instead of treating it as isolated behavior.

I compiled those insights into focused reports and shared them with DSP Owners. The value was not just in calling out violations, but in tying them directly to specific routes, drivers, and consequences. The reports highlighted repeat non-compliance and clearly showed how seat belt violations impacted overall safety scores. Because those scores directly influence bonuses, contract renewals, and route growth, the findings carried real weight. Once the data connected behavior to business impact, DSP Owners responded quickly.

In response, many DSPs implemented targeted coaching to address the behavior, reinforcing that safety protocols applied even on short, repetitive trips. Some went further, requiring DAs to review safety reminders before starting routes in gated communities. These adjustments were tailored to the realities of those neighborhoods, and they worked. Follow-up audits showed a clear reduction in seat belt violations, confirming that focused intervention could change behavior.

That experience reinforced the real value of data when it is applied with intent. By connecting telemetry, routing patterns, and what was actually happening on the road, I was able to surface a problem that had been hiding in plain sight. Working alongside DSPs, we turned those insights into practical changes that improved compliance and aligned safety expectations with real operating conditions. It was a clear reminder that data only matters when it drives action, and that action only lasts when it reflects how the work is actually done.

Compliance, however, was not linear. As monitoring tightened and behavior shifted, a new workaround emerged. Some DAs began buckling the seat belt behind them and sitting on it while driving short distances. The alert system stayed quiet, but the risk did not disappear. It was a behavior shaped by convenience and pressure, and it came to light through continued data analysis and direct observation on the ground.

Once identified, I worked closely with DSPs to address it directly. The data showed how frequently it occurred and how significantly it affected safety scores and operational standing. More importantly, it highlighted the risk drivers were taking and the exposure it created for their businesses. DSPs moved quickly. Coaching was reinforced. Some introduced random compliance checks on gated community routes. Others added manual vehicle checks before dispatch to ensure seat belts were used properly from the start. Accountability increased, and the behavior declined.

This cycle made one thing clear. Data does not just reveal problems. It drives solutions when paired with vigilance and collaboration. By staying ahead of evolving behaviors and adjusting our approach in real time, we were able to address the issue without escalating it into something more serious.

The results were measurable. DSPs could identify individual drivers who needed additional coaching while reinforcing team-wide accountability. Over time, sustained monitoring and targeted intervention led to a significant decline in violations. Fleet safety improved, drivers were better protected, and the communities they served were safer as a result. It was proof that a data-driven, collaborative approach can produce real, lasting improvements in operational safety.

Fleet safety monitoring at Amazon is not a side function. It is built directly into how the operation runs, with the stated goal of protecting Delivery Associates, keeping vehicles roadworthy, and sustaining performance at scale. That process starts before a van ever leaves the yard. Every vehicle goes through mandatory pre-trip and post-trip inspections, documented through the Daily Vehicle Condition Report. These checks are designed to catch issues like tire wear, brake problems, or mechanical faults before they become safety incidents. When a problem is flagged, the vehicle is grounded immediately and does not return to service until the issue is resolved. In practice, this removes risk upstream instead of reacting to failures after the fact.

That foundation is reinforced by a set of monitoring tools that provide constant visibility into both driver behavior and vehicle health. The Mentor app tracks driving patterns and assigns a FICO Safe Driving Score based on behaviors such as harsh braking, speeding, distraction, and aggressive acceleration. Fleet Flash aggregates fleet-level data, highlighting overdue maintenance, grounded vehicles, and broader station health indicators. Together, these tools give managers a real-time picture of what is happening on the road and in the yard. Maintenance compliance is enforced through the Xcelerate Portal, which tracks preventive maintenance intervals and automatically grounds any vehicle overdue by more than five hundred miles. There is no discretion in that process. If the maintenance is late, the van does not move.

Framed together, these systems create a safety net that is both preventative and enforceable. Vehicles are monitored continuously, driver behavior is measured objectively, and lapses carry immediate operational consequences. Whatever criticisms exist around other parts of the operation, the fleet safety infrastructure itself is deliberate, data-driven, and designed to catch problems before they turn into injuries or failures.

Beyond vehicle monitoring, Amazon puts significant weight on shaping driver behavior through constant coaching and compliance checks. Drivers with low FICO scores are flagged quickly and placed into corrective action plans focused on changing habits, not just documenting failures. Other behaviors, like seat belt usage and reversing compliance, are tracked independently to ensure they are not lost inside broader driving metrics. That dual focus matters. Vehicles are only as safe as the people operating them, and Amazon's system treats both as inseparable parts of the same risk equation.

When vehicle damage occurs, it is categorized clearly, whether it is routine wear or major safety-impacting damage, and each category carries defined repair timelines. Any grounded vehicle must pass multiple DVCR audits before it is cleared to return to service. There are no shortcuts. Scorecards and formal failure review reports are reviewed regularly, not just to explain missed metrics, but to identify patterns, assign corrective actions, and prevent the same failures from repeating. Accountability is built into the process, not added after the fact.

Training ties all of it together. Safety is reinforced through recurring sessions designed to keep best practices fresh and encourage proactive problem solving rather than reactive discipline. Combined with continuous feedback and data-driven insight, teams are able to anticipate risks instead of responding to incidents after they happen. When the system works as intended, technology, accountability, and training reinforce each other. The result is an operation that not only protects drivers and the public, but also runs more consistently and efficiently because fewer things go wrong in the first place.

That is why the criticism has always felt incomplete to me. It tends to flatten a complex system into a simple narrative of neglect, when what I saw every day was something far more nuanced. Amazon invested heavily in safety infrastructure, data, training, and enforcement. The machinery was there, and in many cases, it worked exactly as designed. Risks were identified early, behaviors were corrected, vehicles were grounded without hesitation, and patterns were addressed before they escalated into serious incidents. From the inside, it was impossible to argue that safety was ignored or treated casually.

Where the story breaks down is not in the existence of safety systems, but in how those systems were sometimes overridden, distorted, or misused by leadership decisions disconnected from execution. Safety became both a genuine priority and a convenient shield. When leaders made poor process choices, safety language was often used to justify them after the fact, even when those choices introduced new stress, confusion, or risk on the floor. In those moments, the problem was not a lack of safety focus. It was a failure of judgment, humility, and willingness to listen to the people who actually understood how the work was being done.

That distinction matters. A strong safety framework can only protect people if it is paired with leadership that respects operational reality. When frontline insight is ignored, even the best designed systems start to fray. People adapt, cut corners, or invent workarounds, not because they are reckless, but because they are trying to survive inside constraints that no longer make sense. Blaming those outcomes on workers misses the point entirely. The real failure lives higher up, where decisions are made without accountability for their downstream impact.

In the end, what I saw at Amazon was not a company indifferent to safety, but one struggling to reconcile its safety ideals with a leadership culture that too often prioritized optics, speed, and control over listening and learning. When safety was treated as a living practice, shaped by data and grounded in reality, it worked. When it was reduced to a talking point or a justification for bad decisions, it failed the very people it was supposed to protect.

That is the story that rarely gets told. Safety at Amazon was real, serious, and deeply embedded. But like everything else in the operation, its effectiveness depended entirely on whether leaders were willing to engage with the truth on the floor. When they did, people were safer. When they did not, no amount of technology, training, or policy language could compensate.

Respect Is Earned on the Floor

Looking back, one of the quieter strengths of Amazon's operation was its commitment to diversity and inclusion on the floor. It was not framed as something abstract or aspirational. It was simply how the work got done. The warehouse brought together people from vastly different backgrounds, each showing up with their own responsibilities, pressures, and motivations, yet all contributing to the same outcome. College students worked alongside single parents. Retirees shared shifts with people just starting out. The operation functioned because that mix existed, not in spite of it.

By that point in my career, working alongside members of the LGBTQ community was not something that stood out as novel, but it did reinforce a pattern I saw repeatedly. Some of the most consistent, dependable performers came from that community. They took ownership, showed up prepared, and handled pressure without drama. Their contributions were not exceptional because of who they were, but because of how they worked, and the operation benefited from that reliability every day.

What stayed with me was the resilience many of them carried. The ability to navigate challenges outside of work without letting it bleed into performance is not something you can train. It shows up in how people respond to stress, support teammates, and keep moving forward when conditions are not ideal. Over time, that experience stripped the concept of diversity of any abstraction for me. It was not a value statement. It was an operational reality. When people with different life experiences are allowed to contribute fully, the system becomes stronger, steadier, and better equipped to absorb pressure.

Early in my time at Amazon, I was promoted into an Ambassador role, a position that carried real respect on the floor and was not handed out lightly. Ambassadors were expected to understand the operation end to end and lead by example, often with only one assigned per shift. The role mattered. We trained new hires, stabilized the floor when things went sideways, and supported experienced associates when the operation needed direction.

For me, the value of the role was never the title. It was the opportunity to pass on what I had learned and support the people around me. Helping new hires find their footing and watching their confidence grow was rewarding in its own right. Being an Ambassador meant contributing to the culture on the floor and helping the operation succeed as a team.

One day on the warehouse floor, I noticed an associate named Rikky struggling while loading forty-plus-pound totes onto baker's racks. The work was repetitive and physically demanding, and it showed in the way she moved. Once loaded, the racks had to be wheeled to staging for the morning loadout, and the strain on her back was obvious. I stepped in and showed her a more efficient way to grip the tote, position it on the rack, and swing it into place. I also demonstrated a small adjustment in how she pushed and guided the loaded rack, explaining how shifting her hand placement and posture could significantly reduce the stress on her back.

At the end of the shift, Rikky came over to thank me and said the adjustments were already helping. Over the next few months, we worked alongside each other often and developed an easy rapport. One day, almost casually, she mentioned that she was a lesbian. That detail itself did not matter to me. What stayed with me was what she said next. Her wife wanted to thank me for stepping in and helping, and for showing her a way to work that had significantly eased the strain on her back. It was a small exchange, but it landed in a meaningful way, a reminder that support on the floor can carry weight far beyond the task itself.

We continued to cross paths after that, and each time she made a point to thank me again. It was never just about the technique. It was about being noticed in a moment of struggle and feeling supported rather than dismissed. That experience reinforced something I have come to believe deeply. Stepping in when someone is struggling, even in small ways, matters. In a high-pressure environment like Amazon, those moments reveal the human side of the work and the quiet impact we can have on each other when we choose to pay attention.

Not all of my experiences at Amazon were as straightforward as the one with Rikky. During my time at the Portland station, I worked alongside one of the most difficult personalities I encountered in the DSP network, Mel, a DSP Manager whose reputation was well established long before I ever met her. She was known for her aggressive communication style and openly confrontational approach with station leadership. Her presence alone made people tense, and more than a few leaders quietly warned me to keep my distance whenever possible.

What struck me even more than her demeanor was how leadership talked about her when she was not around. They used a derogatory label to describe her, one rooted in stereotypes about lesbian women who present more masculinely. At the time, I did not fully understand the term, but it quickly became clear what it represented. It was not shorthand for difficult behavior. It was bias, casually deployed. That language said far more about the people using it than it did about Mel, and it exposed an undercurrent that still existed beneath Amazon's polished messaging around inclusion.

Avoiding her, however, was never a realistic option. As a DSP Manager, Mel was deeply embedded in the loadout process, and our roles intersected constantly. Every morning required tight coordination to ensure hundreds of totes were staged correctly and loaded into the right vans under unforgiving time constraints. Whatever personal opinions leadership held; the operation depended on functional collaboration. And whether they liked it or not, Mel was a critical part of making that happen.

The work at that station was especially demanding because Amazon was still experimenting with how best to load different types of delivery vans. The trial-and-error nature of those decisions often led to vans being overpacked, with extra totes left behind when space ran out. We called those failures "CubeOuts." More often than not, Mel's routes absorbed the worst of it, routinely receiving more totes than her vans could physically handle. That imbalance threatened on-time delivery and quickly became one of the most persistent operational problems I had to manage.

When CubeOuts occurred, I redistributed the excess totes to nearby routes within the same DSP that still had available space. It was the fastest way to protect customer promises and keep the operation moving. Operationally, it made sense. Interpersonally, it did not go over well. Mel resented the constant reshuffling of her routes and made her frustration very clear. She was vocal, confrontational, and unapologetic about it. I stayed focused on the larger objective. Customer outcomes came first, and keeping the operation functional required coordination, not territorial disputes. No matter how heated the reaction, I kept the conversation anchored in that reality.

Over time, I began to notice a shift in our interactions. Mel would still tell me to "fuck off" now and then, but she followed through with what needed to happen. That alone marked progress. Gradually, the constant friction softened into something closer to mutual respect. We even started laughing about Amazon's perpetually broken routing logic, finding humor in the same chaos that had once fueled our clashes.

As the tension eased, her tough exterior cracked. Beneath it was someone far more approachable than her reputation suggested. One day, without warning, she brought me lunch. It was a small gesture, but an unexpected one, and it said more than any apology ever could.

The relationship took an even more surprising turn when Mel offered to take me on a daylong ride through Oregon on the back of her massive Harley. It was an invitation that said everything about how far we had come. Despite coming from very different worlds, that unlikely connection became one of the most memorable and human parts of my time at the station.

Looking back, I find myself missing Mel and the camaraderie we built through long days and constant problem solving. What began as a tense, adversarial working relationship slowly turned into something genuine. It was a reminder that persistence, shared challenges, and a willingness to look past first impressions can break down barriers that initially feel immovable.

Another colleague who left a lasting impression during my time at Amazon was Jordan, a former military service member who brought an unmistakable intensity to the floor. When something needed to get done quickly and correctly, she was the person I turned to. Her approach was direct and disciplined. She stayed focused, held people accountable, and did not tolerate half effort. Her military background showed in the way she carried herself and in the standard, she expected from those around her. That consistency raised the performance of the entire shift. I trusted her judgment, which is why I often shared operational insights and loadout optimizations with her, knowing she would apply them immediately and without hesitation.

When Jordan told me she was transferring to a station closer to home, it was bittersweet. The move made sense for her life, but it left a real gap on the team. On her last day, she thanked me for the support and training we had shared and said she would miss working together. The feeling was mutual. Her presence had strengthened the operation and pushed the people around her to be better, simply by example.

Her departure was another reminder of how transient the Amazon workforce could be. People moved on for practical reasons, often quickly, but the impact they made lingered. Jordan's influence on the team did not disappear when she left. Neither did the respect we had built working side by side. In an environment defined by pace and pressure, those connections were real, and they stayed with me long after the shift ended.

Looking back on my time at Amazon, one thing became consistently clear. I built strong, positive working relationships with many of the women I worked alongside who identified as lesbian. Those connections were never about identity alone. They were built through shared pressure, mutual respect, and a common understanding of what it took to survive and perform in a demanding environment. Whatever the reason, those relationships became some of the most rewarding and human parts of my experience.

What made that contrast sharper was how uneven leadership often was in navigating the same space. While the work itself fostered collaboration and respect, leadership frequently seemed uncomfortable or ill-equipped to support LGBTQ employees in a meaningful way. In my view, it came down to a lack of maturity and awareness rather than outright hostility. Too often, leadership reduced people to labels or stereotypes instead of recognizing the discipline, resilience, and perspective they brought to the operation. It was frustrating to watch a company that prided itself on innovation fall short in such a basic area, especially when the value of inclusion was already proving itself on the floor every day.

My journey at Amazon included many positive, meaningful experiences working alongside women who identified as lesbian. What stood out was not a policy or a stated commitment, but the consistency of their performance and the strength they brought to the operation. In role after role, I saw work ethic, discipline, and accountability translate directly into results on the floor. That reality mattered far more than any corporate messaging.

From my time as an Ambassador mentoring new hires, to navigating difficult dynamics with strong personalities like Mel, to relying on disciplined associates like Jordan, one pattern became impossible to ignore. Teams succeed when people are judged by how they show up, how they perform, and how they treat others. Respect was earned through action, not identity. Every person contributed something real to the operation, whether it was resilience under pressure, problem solving, or the ability to steady a shift when things started to unravel.

Those experiences reshaped how I think about leadership. Strong teams are built by treating people as individuals and holding everyone to the same standards. Real leadership is not about labels or slogans. It is about paying attention to who carries the load, who can be trusted when conditions get hard, and who consistently does the work without needing recognition to do it. When people are treated with respect and given a fair opportunity to contribute, performance improves naturally.

In the end, my time at Amazon was never just about quotas or metrics. It was about the people I worked alongside, the lessons learned under pressure, and the relationships built in the middle of demanding days. The strongest operations I experienced were the ones where differences existed, but performance was the common language. Not because anyone was elevated or excused, but because everyone was allowed and expected to bring their full capability to the work. That understanding continues to guide how I build trust, teams, and relationships today.

What all of these experiences ultimately taught me was that inclusion does not live in policy statements or internal announcements. It lives in the day-to-day moments where people either feel seen or ignored, supported or dismissed. On the floor, under real pressure, identity fell away quickly and was replaced by something far more honest. Could you be trusted when things went sideways. Could you carry your share of the load. Could you treat the people next to you with respect when the clock was against you. Time and again, the people who answered those questions well became anchors for the operation, regardless of how they identified.

That contrast between the floor and leadership stayed with me. The work itself had a way of leveling people, forcing collaboration and respect through shared necessity. Leadership, by comparison, often lagged behind that reality, clinging to shallow narratives and outdated assumptions while the operation quietly proved them wrong every day. The most effective teams I worked with were not built through messaging or mandates. They formed organically around competence, consistency, and mutual reliance. Inclusion happened not because it was declared, but because it was required to survive the work.

In the end, those relationships shaped me far more than any metric or title ever did. They reinforced a simple truth that has guided me ever since. Strong operations are built by people who show up, do the work, and support one another when it counts. When leaders learn to recognize that reality and get out of the way, teams thrive. When they do not, the work still gets done, but the cost is higher and the talent quietly slips through their fingers. My time at Amazon made one thing clear. Performance is the great equalizer, and respect earned on the floor carries more weight than anything written on a wall.

When Integrity Triggers Hostility and Bullying

To understand how I responded to the conflict that followed, it is important to understand the moral framework I carried into the workplace. My faith shaped how I believed people should be treated, especially when pressure, anger, or self-interest made retaliation seem easier than restraint. The Bible commands us to love our neighbors as ourselves and to treat others as we would want to be treated. I carried that standard into work not as a declaration of moral superiority or an expectation of perfection, but as a way of measuring my own conduct when the environment around me often rewarded something very different.

The relationship with my manager deteriorated over time, although I cannot state with certainty what caused the change. Several disagreements may have contributed. During Holiday Peak, I was asked to complete vehicle-safety records in a way that I believed would overstate what had actually been inspected. I declined because I was not willing to certify work, I could not personally verify. On another occasion, I was encouraged to submit negative documentation about a high-performing colleague, but I did not believe the circumstances justified the record I was being asked to create. I refused that request as well. I had also trained this manager when he first arrived at the station, which added another layer to a relationship that had become increasingly strained, but I cannot know how he interpreted my experience, independence, or decisions.

What I do know is that our professional relationship changed significantly after those disagreements. Conversations that had once been collaborative became colder and more confrontational. Information I previously would have received stopped reaching me, support became less dependable, and decisions affecting my work were increasingly made without my participation. I cannot prove that every exclusion was intentional or that each decision came from the same source, but the cumulative effect was unmistakable. I was operating with less information, less access, and less opportunity to contribute, even though the expectations placed on me had not diminished.

The change eventually extended beyond my direct interactions with one manager. I began to feel increasingly isolated from other station leaders and less connected to relationships I had built across the region. Meetings occurred without me, conversations shifted when I entered the room, and opportunities to contribute seemed to narrow without explanation. I was not present for private discussions and cannot say what was communicated about me, but I could see the effect on my professional standing. Doors that had once appeared open became harder to access, and colleagues who had previously engaged with me freely became more guarded. Whether those changes resulted from direct influence, informal impressions, or broader organizational dynamics, the result was the same: my ability to operate effectively was steadily reduced.

By that point, the environment had moved beyond an ordinary disagreement over performance or leadership style. I experienced each day as a test of endurance, marked by heightened scrutiny, repeated second-guessing, and a growing sense that my judgment was no longer trusted. The psychological toll was real. The sense of purpose that had once anchored my work was gradually replaced by anxiety, exhaustion, and uncertainty about what criticism or documentation might come next. I cannot assign a private motive to the people involved, but I can describe the effect their decisions had on me. Integrity, which I had always considered an essential part of leadership, increasingly felt like something that carried professional consequences.

My career progression also began to stall. Attempts to discuss advancement produced vague feedback, shifting timelines, and conversations that rarely led to a clear path forward. At different points, I was told to keep demonstrating readiness, continue building visibility, or wait for the right opportunity, but the criteria seemed to move each time I met the previous expectation. The process became increasingly difficult to understand because the performance standards I could measure no longer seemed connected to the decisions being made about my future.

I cannot prove that a particular person blocked every opportunity or that each delay resulted from retaliation. There were too many people involved and too many decisions made behind closed doors for me to make that claim responsibly. What I could observe was a widening gap between my operational record and the opportunities available to me. I had delivered measurable improvements, solved problems others had struggled to fix, supported successful launches, developed people, and repeatedly taken on responsibilities beyond my formal role. Yet those results no longer seemed to translate into meaningful consideration for advancement.

Over time, promotion began to feel less connected to measurable performance and more dependent on whether senior leaders trusted, supported, or personally advocated for you. Once certain leaders had formed an unfavorable view of me, strong results did not appear sufficient to change that perception. The work still mattered when Amazon needed a problem solved, but it seemed to matter far less when decisions were being made about who deserved to move forward. That distinction became increasingly difficult to ignore.

Once I concluded that the station environment was unlikely to improve, I began pursuing positions elsewhere in the organization. Several opportunities progressed far enough that I believed I was a credible candidate, but one by one, they ended without a clear explanation. I was not given access to the discussions behind those decisions, so I cannot state that anyone interfered with the selection processes. The pattern was nevertheless difficult to ignore. Roles that appeared aligned with my background would move forward and then stop, often without substantive feedback that might have helped me understand or correct a legitimate concern.

One position was especially disappointing because it matched my technical and analytical background closely. I had genuine interest in the financial analyst role and believed my education and operational experience made me well qualified. The process appeared to be moving forward until a senior regional leader became involved, after which the opportunity ended without an explanation or final conversation. I cannot say what the leader communicated, whether other candidates were stronger, or whether unrelated business considerations changed the decision. What remained with me was the silence and the absence of any feedback that would explain why an opportunity that had appeared promising disappeared so abruptly.

The deterioration had begun months earlier with formal discipline. In April, I received a written warning for failing to escalate a reported altercation between an associate and a delivery driver. I had not witnessed the incident and, to the best of my recollection, no one had informed me about it before the documentation was issued. I was troubled that the record appeared in my file without what I considered a meaningful opportunity to explain that I had no prior knowledge of the event. I suspected that my manager may have influenced the decision, but I did not have access to the full process and cannot establish who initiated or approved it. What I can say is that the warning changed how I viewed subsequent disciplinary actions because it created a formal record based on a failure to report something I did not know had occurred.

The relationship became more openly demeaning in May when my manager referred to me as the “Old Man” in front of others. The comment was presented with humor, but I experienced it as a reduction of my experience and authority to a joke about my age. It was repeated on another occasion, which made it harder to dismiss as an isolated remark or an accidental choice of words. I cannot speak for how everyone else understood the comments, but I knew how they affected me. They communicated that disrespect could be expressed publicly while remaining just ambiguous enough to be defended as humor.

Amazon required managers to conduct regular one-on-one meetings with their direct reports. In my earlier roles, those conversations had generally been constructive forums for discussing operations, professional development, and career goals, followed by written summaries that reflected the major points covered. After this manager assumed responsibility for me, the meetings became weekly and took on a very different tone. I experienced them as increasingly confrontational, dismissive, and focused on criticism rather than development. My decisions were repeatedly questioned, and comments were sometimes made that I considered personal rather than operational. The written follow-ups, however, were much more restrained and did not include the tone or remarks I remembered from the meetings.

That difference between the spoken conversations and the written summaries created another layer of uncertainty. The official record presented controlled, professional exchanges, while my experience inside the room was far more adversarial. I cannot say whether the summaries were intentionally incomplete or simply reflected my manager's interpretation of the discussions. I can say that they did not capture the atmosphere I experienced or the effect those meetings had on me. As the documentation accumulated, I became increasingly concerned that the written record would be treated as the complete history while the conduct that shaped the meetings remained invisible.

In June, I was told that I would receive greater involvement in the business side of the operation. I welcomed the statement because it suggested that my experience might still be valued and that the relationship could be reset through meaningful work. I prepared to participate, but the promised meetings did not occur, responsibilities were not assigned, and my follow-up questions did not produce a clear path forward. Strategic discussions continued without me, and peers occasionally asked why I was not participating. I cannot say that my absence was intended to damage my reputation, but it created the appearance that I was either uninvolved or no longer considered relevant to work I had previously supported.

Viewed individually, any one of these events might have been explained as a communication failure, a management preference, or an ordinary business decision. Viewed together, they formed a pattern that was much harder for me to dismiss. I had received discipline for an incident I did not know about, been subjected to demeaning comments, excluded from meetings, denied promised responsibilities, and left without meaningful answers about career opportunities. I cannot prove that these actions were coordinated or undertaken with the specific goal of forcing me to resign. What I can say is that their combined effect steadily eroded my standing, confidence, and ability to see a future in the organization.

The clearest acknowledgment came in July during a conversation with a new manager. He described me as a "square peg trying to fit into a round hole" and questioned why someone with my operational experience and knowledge had not advanced. He also said that my former manager had placed what he called a "dark cloud" over my career. Those were his words as I remember them, and I understood them as confirmation that the concerns surrounding me extended beyond my own perception. His comments did not prove that every missed promotion, disciplinary action, or lost opportunity resulted from deliberate interference, but they gave me a clearer explanation for why my performance and experience had not translated into advancement.

That conversation stripped away much of the uncertainty I had been carrying. I had spent months asking whether I had misread the environment, failed to communicate effectively, or allowed frustration to distort my interpretation of events.

Hearing another manager acknowledge that a negative perception had followed me made it harder to believe that the problem could be solved through performance alone. The system was not necessarily operating through a formal decision to end my career, but it had allowed an unfavorable narrative to become influential enough to shape how future opportunities were viewed. Once that narrative became embedded, correcting it appeared far more difficult than creating it had been.

The experience forced me to confront a deeper tension between the principles Amazon promoted and the behavior I had encountered. The company spoke often about ownership, earning trust, and having the backbone to disagree and commit. I believed I had followed those principles when I declined to certify safety work I could not verify and refused to create negative documentation I did not believe was justified. I cannot claim that my interpretation of every request was the only possible interpretation, but I acted according to my understanding of the facts and my responsibility as a leader. Instead of strengthening my standing, those decisions appeared to contribute to a relationship in which my judgment was increasingly viewed as resistance rather than integrity.

From a faith perspective, the conflict cut even deeper. Loving your neighbor and treating others as you would want to be treated are not abstract ideals when authority is involved. They require restraint when retaliation is possible, fairness when documentation can damage a career, and honesty when metrics create pressure to present events more favorably than they occurred. What disturbed me was not simply that leaders could make poor decisions. It was that a system built around principles and accountability could allow people to be reduced to obstacles, liabilities, or narratives that needed to be managed.

By the time the new manager spoke openly, I no longer believed that staying represented perseverance. It had begun to feel like a gradual surrender of my confidence, health, and sense of professional identity. I could not prove that the organization had collectively decided to force me out, but I could see that the environment was not correcting itself and that the opportunities available to me were continuing to narrow. No amount of patience could change a system in which a damaging perception had already become more influential than the results I had produced.

What remained was clarity rather than vindication. I had not handled every moment perfectly, and I did not possess every fact behind every decision. I did know that I had refused to sign records I could not verify, declined to document a colleague in a way I considered unsupported, and continued performing in an environment that had become increasingly difficult to navigate. I also knew that remaining indefinitely would have required me to ignore what the experience was doing to me.

That chapter ended without an apology, a corrected record, or a promotion that restored what had been lost. It ended with a clearer understanding of who I was, what I had tried to protect, and what I was unwilling to become in order to advance. Titles and promotions mattered, but they mattered less than leaving with my values intact. In the end, that was the one outcome the system could not take from me.

HR Was Never a Safe Place

By the time I turned to HR, the damage was already taking a toll, but I still believed the system existed to protect people like me. The psychological strain was cumulative and corrosive. Each day required a level of vigilance that had nothing to do with the work itself. I measured my words, anticipated backlash, and replayed conversations long after they ended. Sleep grew lighter. Anxiety followed me home. The job was no longer something I did. It was something I endured.

What made it worse was the isolation. The systems meant to provide balance, feedback, and protection were either compromised or silent. I was navigating hostility alone, fully aware that the person creating the environment held increasing power over my future.

At some point, the question shifted. It was no longer how I could fix the situation, improve alignment, or wait it out. The question became how much more damage I was willing to absorb. Staying no longer felt like resilience. It felt like self-betrayal. I realized that continuing to operate inside a system designed to break me would only cost more of what I could not afford to lose: my confidence, my health, and my sense of self. That was the moment it became clear. Survival did not mean enduring longer. It meant leaving before the damage became permanent.

August brought further strain, both physical and emotional. Years of walking concrete warehouse floors finally caught up with me, and an ankle injury left me hobbling through the station as I struggled to keep up with my responsibilities. There was no concern, no check in, no acknowledgment of the physical toll the work had taken. Instead, during a SUM in front of more than sixty associates, Mark turned my condition into a spectacle. He joked that my "old body" was breaking down and suggested I get a motorized wheelchair. It was not offhand or careless. It was public, calculated, and meant to humiliate. The message was clear. My pain was entertainment, and my dignity was expendable.

In August, the pattern crossed another line. Mark told me that Ken, an older colleague, had expressed interest in a driver trainer role. He then instructed me to prevent Ken from pursuing the opportunity. The directive felt wrong immediately. It was inappropriate, discriminatory in tone, and revealing of a mindset that viewed age and longevity as liabilities rather than assets. It reinforced what I had already come to understand. This was not isolated behavior. It was a worldview.

By early October, the hostility became impossible to ignore. Seth informed me he was resigning, telling me directly that he had reached his limit with the toxic culture. Before leaving, he reiterated his belief that Mark was the primary obstacle to my growth and that Mark was actively generating hostility among station leadership toward me. Hearing it again, from someone with nothing left to gain, carried weight.

Later that same day, I called Jerry, a senior regional manager I had worked with previously, to voice my frustration. He listened carefully and acknowledged my experience, describing me as a senior operations manager with significant expertise and real value. He told me he would speak with Mark and follow up. Like so many assurances before it, nothing came of it. There was no follow up. No intervention. Just silence. By that point, the pattern was complete. The behavior was known. The impact was acknowledged. And still, nothing changed.

October brought a sharp escalation in personal attacks that further poisoned an already toxic environment. Mark mocked the new shoes I wore to manage my worsening ankle pain, derisively calling them "circus shoes." It was petty, unnecessary, and meant to embarrass. Later that same day, he quietly criticized an ad hoc process and disparaged my performance in a conversation with Eric, a colleague. To Eric's credit, he pushed back, stating he was unaware of any issues with my work. That moment only underscored the pattern. The criticism existed whether the facts supported it or not.

The hostility escalated further in mid-October during a manager meeting, when Mark openly declared that I was "too old to work at Amazon." There was no attempt at humor or ambiguity. The remark landed as it was intended, dismissive, ageist, and humiliating. By that point, the behavior was no longer subtle or deniable. It was openly hostile and deeply demoralizing, making it increasingly difficult to maintain focus or morale in an environment that seemed determined to break me down.

By late October, the pressure turned overtly punitive. Mark ambushed me with baseless accusations, blaming me for poor station performance metrics, PNOV, OODT, and BC, despite those metrics clearly falling under the responsibility of the RTS team, not mine. He promised to provide a list of issues for me to address. As expected, that list never came. The accusation stood without evidence, clarification, or follow-up, leaving me in a constant state of uncertainty and defense.

That same day, Mark disregarded standard expectations and demanded that I work beyond my ten-hour shift. The pressure did not ease. By the end of October, he began messaging me on my days off, asking for operational reports that were the responsibility of on-shift managers, not mine. There were no boundaries, no respect for time off, and no pause. The behavior felt relentless, designed to keep me constantly off balance and under scrutiny.

November marked another escalation. During a one-on-one meeting, Mark ridiculed my performance, accused me of failing to follow station processes, and again demanded that I work during my off hours. When I asked for clarification or a written outline of expectations, his demeanor turned openly antagonistic. The request itself seemed to provoke him. Rather than provide direction, he escalated the confrontation, increasing the tension and reinforcing that the goal was not improvement, but pressure.

By mid-November, another one-on-one meeting quickly devolved into the same pattern of criticism. Mark chastised me for delayed emails, not reviewing documents with PAs, and missing routes, issues that were either misrepresented or outside my direct control. Once again, he reiterated his expectation that I work on my own time. When I attempted to explain my perspective, he dismissed it outright, leaving no room for discussion or resolution.

Later that same day, Mark approached Michael, another colleague, to question my performance, clearly attempting to build support for further action against me. Michael confirmed there were no performance issues and told Mark directly that I was being deliberately bypassed in the sharing of process-related information. That confirmation only reinforced what had become painfully obvious. This was no longer about accountability or results. It was a calculated effort to isolate, discredit, and push me out, carried out through sustained pressure, misrepresentation, and intimidation in an environment that had become increasingly hostile and inescapable.

By that point, it was clear the situation would not resolve itself. Desperate for a way out, I reached out to Elena, my Senior HR Business Partner, with whom I had built a strong working relationship over the years. I laid everything out: the daily micromanagement, the escalating hostility, and the toxic environment Mark had created and sustained. I shared detailed documentation of each incident; records I had been keeping since the earliest signs of the abuse. Elena expressed genuine concern for my well-being and told me she would help explore options for transitioning into a new role. After our meeting, I sent a thorough follow-up, reiterating the severity of the situation and thanking her for her support. For the first time in a long while, I felt cautiously hopeful.

What I did not realize at the time was that Elena had initiated an internal investigation.

When I learned about it, I had no doubt Mark would be furious. By then, the company was loudly promoting its Diversity, Equity, and Inclusion initiatives, and the environment around promotions and accountability felt increasingly political. Against that backdrop, I was already feeling sidelined and expendable. Still, I resolved to approach the investigation professionally. I organized my notes, aligned timelines, and prepared to walk through the documented incidents calmly, expecting a fair and impartial review.

The investigator's call shattered that expectation almost immediately. Her tone was adversarial from the start. Questions felt leading. Clarifications were brushed aside. The focus seemed less on understanding what had happened and more on framing me as the problem. I hung up the call feeling stunned and deeply demoralized.

Afterward, I looked up the investigator using internal tools that display employee profiles, organizational information, and tenure. I discovered she had been with the company for only three months and had no operational background relevant to the situation she was evaluating. That realization made the outcome feel increasingly predictable. When the investigation concluded, it found no evidence of harassment. The decision ignored my documentation, the corroborating patterns, and the consistency of the behavior I had reported.

The result was not just infuriating. It was disillusioning. The process felt performative, designed to check a box rather than confront a problem. I had followed the rules, documented the abuse, and trusted the system to act in good faith. Instead, the experience left me questioning whether the process existed to protect employees at all, or simply to insulate the organization from accountability.

The final breaking point was not a single comment or meeting. It was the realization that HR was not a safe avenue, not because they were unaware, but because they were unwilling to act. I had done everything the right way. I documented incidents in real time. I escalated professionally. I spoke calmly and factually. I trusted the process I had spent years encouraging others to trust. And in return, the system closed ranks.

The investigation did not fail due to lack of evidence. It failed because acknowledging the truth would have required confronting a rising leader and admitting that the behavior had been tolerated, if not enabled. The moment the report concluded there was "no evidence of harassment," despite months of documented abuse and corroborating patterns, I understood what had really happened. The process was never designed to protect me. It was designed to neutralize risk.

That was the moment the illusion collapsed. HR was not a safeguard. It was a containment mechanism. Once I saw that clearly, there was no path forward inside the organization. Staying no longer meant resilience or professionalism. It meant accepting damage as the cost of employment. I realized then that the safest decision I could make was to remove myself from a system that had proven it would rather preserve appearances than protect people. That was the end. Not of my effort, but of my belief that the institution would ever choose integrity over convenience.

Mark's tactics shifted immediately after the investigation was closed. The hostility became procedural. He began assigning me unrealistic deadlines for operational reports, flooding me with unrelated training modules that carried impossible completion windows, and scheduling excessive, low value meetings that fractured my workday. None of it was accidental. The intent was clear. Overload me, destabilize me, exhaust me, and force me to quit without ever having to say it out loud.

Quitting was not something I was willing to do. Years of navigating difficult environments had hardened my resilience, and I refused to hand him that outcome. But endurance has limits. The constant pressure, the lack of boundaries, and the knowledge that the system would not intervene took a real psychological toll. I reached a point where survival meant stepping back. To protect my mental health and regain clarity, I made the decision to take a leave of absence and consult legal counsel, preparing myself for whatever came next.

My legal team found what HR would not acknowledge. They identified clear evidence of workplace harassment and age discrimination embedded in Mark's actions. The pattern was familiar to them, one they had seen repeatedly in large organizations quietly pushing out older, more experienced employees in favor of younger, less seasoned hires. Suddenly, the stalled promotions made sense. Managers I had trained, some with a fraction of my experience, were advancing while I was systematically sidelined.

I had documentation. I had corroboration. I had a case. And yet, I hesitated. Amazon's legal machine appeared to me to be built to overwhelm, not resolve. I knew this firsthand. I had previously supported Amazon's legal team and had seen how litigation was used as a war of attrition, burying opposing counsel under endless filings until the cost became unbearable. Pursuing justice meant bracing for a fight designed to drain time, money, and energy.

In the end, I chose not to file. Instead, I returned to work fully aware of what I was walking back into. The decision was not rooted in confidence in the system, but in a calculated assessment of risk and cost. The experience left me deeply disillusioned. Any remaining faith I had in institutional protection was gone. What remained was resolve. I would continue to conduct myself with professionalism and restraint, even inside an environment that had proven itself hostile, biased, and stacked against me.

What I learned was simple and devastating in my experience. When integrity threatened power in my experience, institutions did not correct the abuse. They managed the risk.

That lesson changed how I understood power, accountability, and the stories organizations tell about themselves. Amazon spoke fluently about values, leadership principles, and ethical standards, but in my experience, when those ideals collided with convenience, hierarchy, and self-protection, they folded quickly. Integrity was celebrated in theory and punished in practice, from what I witnessed. The distance between what the company claimed to stand for and how it actually behaved when challenged was no longer abstract to me. I had lived inside that gap.

Returning to work after leave felt surreal. I moved through the station with clarity but without illusion. I did my job well, documented everything, and kept my head down, not out of fear, but out of realism. I understood now that survival inside that system required emotional detachment. Expecting fairness only created more harm. The work became transactional. I gave what was required and guarded what mattered most, my health, my values, and my sense of self.

What stayed with me most was not the cruelty itself, but how normalized it had become. None of what happened required secrecy. The comments were public. The exclusions were visible. The retaliation followed predictable patterns. And yet, because it was carried out by someone ascending the hierarchy, it was tolerated, rationalized, and ultimately erased by process. That realization stripped away any remaining belief that silence was accidental. It was structural.

In the end, the experience clarified something I carry with me now. Systems do not drift into injustice by mistake. They arrive there through repeated decisions to protect power over people. When that becomes clear, the responsibility shifts inward. You stop asking whether the system will do the right thing and start deciding what you are willing to endure, what you will walk away from, and what you refuse to become.

I did not leave Amazon with bitterness. I left with discernment. I knew exactly where the line was, and I knew I had crossed it not by wrongdoing, but by refusing to abandon my principles. That distinction matters. The institution may have closed ranks, but it did not get to define my worth, my integrity, or the truth of what happened. That part was never theirs to take.

If there is a closing truth in all of this, it is this: character is revealed not when values are easy to uphold, but when they carry a cost. I paid that cost knowingly. And while the system moved on unchanged, I did not leave empty-handed. I left with clarity, self-respect, and the unshakable knowledge that walking away from a corrupt dynamic is not failure. It is fidelity to something larger than any title, badge, or employer could ever offer.

A New Demon

Upon my return from leave, the outcome of everything that had come before became clear. Mark was no longer my direct manager. Instead, I now reported to Landon, someone I had previously worked alongside when he held the same level as me. His promotion was not reassuring. It was unsettling.

I had seen Landon's leadership up close. He routinely displayed a lack of emotional intelligence and restraint, berating associates on the sort line over minor mistakes, sometimes reducing them to tears. How someone with that behavior could advance in a company that claimed to be process driven and leadership focused was difficult to reconcile. And yet, it fit a pattern I had already come to recognize. Advancement was not about capability or character. It was about alignment.

It did not take long to understand what this change meant. Landon wasted no time asserting control. His behavior was calculated and intrusive. He would stand directly behind me at my desk while I was deep into critical tasks, interrupting with trivial questions designed to break concentration. The intent was obvious. He was waiting for a mistake, any mistake, that could be used to justify my removal.

What he underestimated was muscle memory. Years of operational experience had made the work instinctive. The processes were second nature, and despite his constant interference, I continued to execute with precision. That only seemed to frustrate him further.

Unfazed by my ability to continue performing, Landon escalated his approach by increasing the frequency of one-on-one meetings. These sessions were not intended to provide guidance or support. They were exercises in sustained criticism. He dissected my daily activities line by line, assigning blame for delays and issues that were either outside my control or unrelated to my role altogether. Each meeting became something to brace for rather than benefit from. Like Mark before him, Landon extended the harassment beyond the workplace, calling me on my days off to demand reports and updates. Boundaries dissolved entirely. The pattern was unmistakable and deliberate, designed to wear me down, undermine my confidence, and push me toward failure. As I had learned to do, I documented every interaction in detail, maintaining a precise log of each incident.

Recognizing that the situation was continuing to escalate, I attempted once more to introduce accountability by requesting the presence of the station's HR representative in these meetings. I hoped a neutral observer might temper the behavior and rebalance the dynamic. Instead, the decision backfired. The HR representative did not act as an impartial mediator. They aligned themselves with Landon, reinforcing his narrative and validating his conduct. The meetings shifted immediately, no longer one-on-ones but coordinated sessions that functioned as joint efforts to further undermine and isolate me. At that point, any remaining illusion of neutrality was gone.

The culmination of the toxic environment came during the annual pay review. Landon and Mark sat across from me and informed me that my performance had been rated as subpar and that I would not be receiving a pay increase. The assessment itself felt hollow, disconnected from both my actual work and the feedback I had been receiving in real time. What stood out most was not the decision, but the delivery. There was an unmistakable satisfaction in the room, a quiet smugness that stripped the moment of any pretense of professionalism. It was clear this was not about performance. It was another calculated attempt to diminish, discourage, and apply financial pressure in hopes that I would remove myself.

The following one on one meeting with Landon and the HR representative took an abrupt and unsettling turn. The tone was suddenly positive. I was given favorable feedback on my performance from the previous week, praise that directly contradicted the narrative used to justify the pay decision. The shift was so stark it was disorienting. I left the meeting cautious rather than relieved, aware that consistency had long since disappeared from the process. By that point, praise no longer felt corrective or sincere. It felt tactical, another move in a pattern that had taught me to question every sudden change in tone.

More importantly, I had scheduled interviews for roles in Seattle and Savannah, positions that offered a genuine chance to escape the environment I had been trapped in for far too long. The prospect of leaving my current station, and more broadly, any warehouse shaped by the same toxic leadership, became a rare source of relief. For the first time in months, there was a visible path forward. My intent to leave was no secret. Management seemed to sense it as well, responding not with support, but with a heightened awareness that their control over me might soon end.

The following day brought a betrayal I did not anticipate. Landon and the HR representative summoned me to a meeting where I was presented with what amounted to a veiled termination, delivered under the framework of Amazon's Pivot program. I was given two options: resign and accept a modest severance, or enter the "Improve at Amazon" program, ostensibly intended to address performance deficiencies. The premise collapsed almost immediately.

Landon claimed my performance failed to meet his standards, but the allegations were demonstrably false. He accused me of failing to complete a critical process on a specific date that fell on a Saturday, a scheduled day off when I was not onsite. He then alleged dropped routes and a significant decline in On Time Delivery on another date, despite the data showing no dropped routes and an OTD of 99.33 percent. Finally, he cited my failure to complete D minus one Capacity Planning SIM, a process for which I had never received training, instruction, or access. Taken together, the meeting stripped away any remaining ambiguity. This was not about performance. It was a coordinated effort to force my exit, carried out through fabricated claims and reinforced by HR's quiet complicity.

Amazon's Pivot program is positioned as a performance improvement process for employees deemed to be falling short of expectations. In theory, it is triggered after prior feedback and coaching fail to produce improvement. Employees are presented with two options: resign with severance or enroll in an improvement plan that outlines specific goals and timelines aligned to Amazon's Leadership Principles. Participants are assigned Career Ambassadors for guidance, and successful completion allows them to continue their careers. If the plan is not successfully completed, employees may either accept severance or pursue an appeal, which is reviewed by a separate manager or panel. On paper, the program is framed as developmental and procedural.

In practice, the program is deeply flawed, particularly when initiated by a manager with a personal agenda. The same leader pushing for an employee's removal is often responsible for defining the expectations, metrics, and evaluations used to judge success, creating an inherent conflict of interest. This structure undermines any claim of objectivity. The support mechanisms offered, including Career Ambassadors, are largely superficial and lack both the authority and context to meaningfully intervene. The severance option is modest, leaving employees with little real choice beyond entering a process that feels designed for failure. Far from serving as a genuine opportunity for improvement, Pivot functions more often as a controlled pathway to dismissal, one that cloaks subjective decisions in procedural language and raises serious questions about fairness and intent.

Ironically, about a year after my own experience, I was asked to serve as a review manager in a Pivot appeal. The role required me to assess the case of an employee outside my reporting chain, ostensibly to provide an unbiased perspective. Almost immediately, the situation felt familiar. The employee was older, accused of underperformance, and cited for failing to follow organizational processes.

As I reviewed the case and interviewed both the employee and their manager, the gaps became impossible to ignore. When I asked the manager to produce a Statement of Work or any formal documentation outlining the processes the employee was allegedly mishandling, the answer was blunt. None existed. There were no written expectations, no defined standards, no documented guidance. The accusation rested entirely on interpretation. The question answered itself. How could someone fail to meet expectations that had never been articulated? The appeal ultimately ruled in the employee's favor.

That experience removed any remaining doubt. Pivot is not a system designed to help employees improve. It is a mechanism that enables removal while preserving the appearance of process. Its reliance on vague standards, shifting expectations, and subjective judgment creates an environment where outcomes are easily manipulated. Managers with ulterior motives are given a tool they can weaponize, particularly against older employees or those willing to challenge decisions rather than comply quietly. Amazon's public language around fairness and high standards collapses under scrutiny in these cases. For a company that prides itself on rigor and operational excellence, the continued reliance on a system so easily exploited is not just disappointing. It is indefensible.

The accusations Landon leveled against me, built on demonstrable lies, falsehoods and distortions, left me stunned. Throughout my career at Amazon, I had consistently delivered strong results and carried significant responsibility. To suddenly be treated as a performance problem, based on fabricated claims used to justify placement on Pivot, was a profound shock. It was not just a professional setback. It was a direct assault on my sense of fairness and trust in the organization. The conclusion became unavoidable. This was not about my work. It was a calculated effort to force me out.

After years of dedication, resilience, and results, the message was unmistakable. Loyalty and performance offered no protection in a system willing to shield poor leadership and tolerate a toxic culture. The sense of betrayal ran deep.

With nothing left to preserve, I made a deliberate decision. I drafted a comprehensive, carefully structured email that documented my accomplishments, detailed the harassment I had endured, and formally challenged the legitimacy of my placement on Pivot. This was not an emotional appeal. It was a factual record supported by performance data, timelines, and documented contradictions that dismantled the narrative being constructed around me. I refused to allow falsehoods to become the permanent version of my career.

Using PhoneTools, Amazon's internal employee directory, I identified the senior HR executive with direct authority over the process, the Senior Vice President of People eXperience and Technology, a direct report to Andy Jassy, CEO of Amazon. I addressed the email to her directly and copied her leadership team to ensure transparency and prevent quiet dismissal. I did so with clear intent. If the system claimed to value rigor, accountability, and leadership principles, then this situation demanded scrutiny at the highest level. I believed that presenting the facts plainly and publicly would force a reckoning, or at the very least, prevent the truth from being buried.

For a brief moment, I allowed myself to believe that this time would be different. At that level of leadership, I expected the facts to matter and the injustice to be acknowledged. That expectation did not last. What followed was another hollow review, one that appeared designed to absorb the complaint rather than examine it. There was no meaningful investigation, no accountability, and no resolution. The process concluded quietly, leaving everything exactly as it had been.

That outcome made one truth impossible to avoid. The culture I had been fighting was not the product of a few bad actors. It was systemic, entrenched, and protected. The behavior was known, tolerated, and ultimately reinforced by leadership structures more concerned with preserving authority than addressing misconduct. The realization was not just disappointing. It was clarifying. A company I had once respected for its rigor and ambition had revealed a different priority altogether, one where maintaining power outweighed fairness, integrity, and the people who carried the work forward.

Preparing for my Pivot review was a grueling process. I meticulously gathered evidence and crafted detailed rebuttals to counter the false claims Landon had made against me. That documentation included screenshots of his Amazon's real-time communication platform app messages that directly contradicted his accusations, text messages sent to my personal phone, and call logs showing repeated contact on my scheduled days off. Even with that preparation, the review itself was a nerve-wracking ordeal, one I would not wish on anyone. I was subjected to intense scrutiny by three review managers and forced to defend actions that had already been misrepresented. In the end, all three ruled in my favor. I successfully passed the Pivot program, formally disproving the allegations that had been used to justify my placement in it.

As outlined in the process, I was then assigned a Talent Advisor to support my transition into a new role with a high-performing manager on a different team. That phase quickly exposed another failure in the system. The Talent Advisor provided little meaningful guidance or support, and efforts to identify suitable roles were minimal at best. What was presented as a structured pathway forward felt like an afterthought, lacking urgency, ownership, or genuine intent. The inefficiency was not incidental. It was emblematic of the Pivot program itself. Even when an employee survives the process, the system offers little real help on the other side, reinforcing the sense that Pivot is designed not to rehabilitate talent, but to quietly move people along once they have already been marked.

Burned out from the sustained stress and emotional toll of the ordeal, I took much-needed vacation time to recover. Even in my absence, the damage continued. While I was away, Landon terminated two of my Process Assistants over minor safety technicalities that, under normal leadership, would have been handled with coaching or corrective guidance. Instead, he went straight to termination. These were not marginal performers. They were exceptional. I had personally identified them from the Associate pool, mentored them closely, and helped them earn their promotions into PA roles. Their operational knowledge, work ethic, and commitment to the team were undeniable. Losing them was a significant blow, not just emotionally, but operationally.

The way it was handled made the intent unmistakable. This wasn't about safety culture or standards. It was punitive. Vindictive. A power move. Watching a fucking little shit like Landon dismantle a team I had spent years building was infuriating, but it was also revealing. The decision showed a complete lack of judgment, leadership, and respect for the people doing the work. It wasn't incompetence alone. It was cruelty disguised as compliance.

What followed made it worse. Despite the destruction left in their wake, both Mark and Landon were promoted into higher leadership roles. That outcome was the final insult. It confirmed what had become impossible to ignore. Performance, integrity, and leadership were secondary to political alignment and internal maneuvering. The promotions were not anomalies. They were proof of how the system actually rewarded behavior.

Disillusionment doesn't begin to cover it. Watching individuals who embodied everything wrong with the culture advance, while capable, committed employees were pushed out or discarded, forced a reckoning. This was a company that spoke relentlessly about leadership principles, innovation, and standards, yet consistently elevated people who violated those values in practice. The cost was real. Talent was wasted. Teams were damaged. And the gap between what Amazon claimed to stand for and how it actually operated had never felt wider.

What stayed with me after all of that was not just the anger or the sense of betrayal. It was the clarity. Once you see how a system actually rewards behavior, you cannot unsee it. Amazon did not fail me because I lacked resilience, skill, or results. It failed because I refused to participate in the quiet compromises that keep toxic hierarchies intact. I would not falsify records. I would not sacrifice other people to protect someone else's metrics. I would not shrink myself to make insecure leaders feel bigger. In a culture that increasingly valued obedience over judgment, that refusal was unforgivable.

The irony is that everything I was punished for was exactly what Amazon once claimed to value. Ownership. Backbone. Doing the right thing when it is uncomfortable. Protecting people while delivering results. Somewhere along the way, those principles stopped being guardrails and became branding. They were still spoken aloud, still printed on walls, but selectively enforced. When principles threatened power, power won. Every time.

By the end, I was no longer confused about why things unfolded the way they did. The pattern was consistent. Experienced operators who questioned bad decisions were sidelined. Older leaders who carried institutional memory were treated as liabilities. High performers who protected their teams instead of playing politics were marked. Meanwhile, those who mastered optics, narrative control, and upward loyalty advanced quickly, regardless of the damage they caused below them. That was not dysfunction at the edges. It was the operating model.

Leaving that environment was not a loss. It was a release. I walked away with my integrity intact, my record unbroken, and my conscience clear. I did not get the promotions I earned or the recognition my work justified, but I kept something far more valuable. I kept my sense of right and wrong. I kept my respect for the people who did the work. And I kept the ability to look back on my career without having to explain away choices I would be ashamed to own.

This chapter is not about revenge or grievance. It is about truth. About how systems that grow as large and powerful as Amazon can lose their moral center while still appearing successful from the outside. About how good people are often pushed out not because they fail, but because they refuse to bend. And about how leadership, real leadership, has nothing to do with title or promotion, and everything to do with who you protect, who you empower, and what you are willing to risk to do the right thing.

Toxic Leadership, Protected Power

Recovering from the mental turmoil of those events took time. I did not bounce back quickly, and I did not return unchanged. After several months, however, I found myself back in the saddle, doing the part of the job I loved most: Morning Dispatch. It was grounding because it was immediate and real. Drivers still needed to leave safely and on time, associates still depended on one another, and people still showed up trying to do the right thing. When ego, politics, and organizational maneuvering fell away, the work itself still mattered. Morning Dispatch reminded me that beneath the leadership failures and corporate noise, there were people trying to keep the operation moving with professionalism and purpose.

By then, two of the leaders connected to the most difficult period at my station had been transferred to other locations. On paper, the moves could be interpreted as action, but from where I stood, they looked more like relocation than resolution. I was not involved in the decisions behind those transfers, so I cannot say what concerns were discussed privately or what expectations accompanied the moves. What troubled me was the absence of any visible reckoning with the damage left behind. The organizational chart changed, but nothing I saw suggested that the leadership practices responsible for so much fear and instability had been confronted. Large organizations can easily mistake movement for progress because a reassignment produces something measurable. A name disappears from one roster and appears on another, allowing everyone to act as though the issue has been handled even when the underlying behavior remains untouched.

Over time, I came to see that pattern as part of a broader cultural weakness. Amazon measured performance relentlessly, but it was far less consistent in examining how those results were achieved. A leader could deliver the expected numbers while leaving people exhausted, distrustful, or afraid to speak, and the system often seemed better equipped to reward the output than evaluate the human cost. Empathy and emotional intelligence appeared secondary to operational performance, especially when the leader involved continued to meet visible expectations. The company knew how to track packages, routes, attendance, defects, and productivity down to the smallest detail, yet the damage caused by poor leadership was much harder to find on a dashboard. When harm could not be easily quantified, it became easier to ignore, explain away, or move to another building.

Before one of those leaders left the station, another event unfolded that stayed with me long afterward. It involved the owner of an independent delivery company operating from the building, someone I had worked with from the beginning of his onboarding process. A serious allegation involving inappropriate workplace conduct was made against him. I was not present for the reported interaction, did not see the original complaint, and did not participate in the investigation. I also did not have access to the complete investigation file, which means I cannot state everything that was alleged or identify all the evidence considered by the decision-makers. What I can describe is the sequence I personally witnessed. The allegation was followed by an internal review, and the delivery company's contract was subsequently terminated. From my limited vantage point, the process appeared to move quickly, and I saw little that gave me confidence that competing evidence had received a meaningful review.

I understand that portions of an internal investigation may take place outside the view of station personnel. Interviews, documents, legal considerations, and additional evidence may have existed that I never saw. Even so, the gap between the seriousness of the outcome and the evidence available to me raised questions I could not dismiss. Surveillance footage existed from the area associated with the reported incident, and I personally viewed footage that I understood to be relevant. Based on what I saw, the recording did not appear to show the kind of conduct I expected given the gravity of the allegation and the severity of the response. The footage did not provide an obvious explanation for why an entire business was about to be dismantled. That observation does not allow me to claim that the recording disproved every possible allegation, because I did not know whether other conduct had been reported, whether something occurred outside the camera's view, or whether investigators possessed additional evidence unavailable to me. I could judge only what was in front of me, and what I saw did not appear to match the outcome that followed.

I had known the owner as a disciplined and professional operator who took his responsibilities seriously. During the time I worked with him, I never personally witnessed him behave inappropriately toward a manager or employee. My experience could not establish what happened in an interaction I did not witness, but it made the allegation difficult to reconcile with the person I had known. His company had also built a strong operational record. It was not a failing contractor drifting toward termination because of missed expectations or poor performance. From everything visible to me, the operation was delivering the results Amazon routinely demanded from its delivery partners. That did not make the owner incapable of wrongdoing, but it did mean the termination could not be easily explained as the predictable end of a struggling business relationship.

There was another fact I could not ignore. Before the allegation arose, the senior leader involved had told me directly that he disliked the owner and wanted the delivery company gone. He did not provide a detailed explanation, and I cannot prove that his statement influenced the investigation or the later contract decision. I can only describe the sequence as I experienced it: the leader expressed a desire to remove the owner, a serious allegation later emerged, an internal process followed, and the contract ended. That sequence is not proof that the situation was orchestrated, but it is why I could not view the termination as a routine or isolated business decision. The earlier statement remained in my mind as I watched the operation collapse. I also understood that the leader occupied a position of influence while the contractor depended heavily on a single corporate relationship for the survival of his business. That imbalance of power shaped how I understood everything that followed.

The consequences were severe and immediate. The owner lost the business he had built around the contract, employees and drivers were affected, and his family faced significant financial uncertainty. His professional reputation also came under a cloud that no later explanation could easily remove. I remained in contact with him during part of that period and saw the strain the situation placed on him. I cannot speak for every private consequence or disclose conversations that were not mine to share, but the effect on him was unmistakable. The dispute eventually entered litigation, and I was later asked to provide information concerning my own experience with the owner. My role was limited to what I had personally observed and the character he had demonstrated in his dealings with me. I was not in a position to determine the ultimate truth of every disputed fact, nor did I pretend to be. By then, however, the damage was no longer theoretical. The contract had ended, the operation had been dismantled, and the people who depended on it had already absorbed the consequences.

What also stayed with me was the apparent absence of consequences for the leader whose earlier statement I had heard. From where I stood, his career continued without visible interruption, and I later learned that he had moved into a role with broader responsibility. The only word that came to mind was unfuckingbelievable, yet the outcome was also consistent with what I had come to understand about the system. Accountability did not necessarily follow the path of harm, and the person or business that absorbed the greatest consequences was not always the person with the greatest influence over the events. I cannot claim to know what senior decision-makers intended or what they discussed behind closed doors. I cannot say what evidence existed beyond the footage and circumstances available to me. What I can say is that the process produced an outcome I could not reconcile with what I had personally seen and heard.

That distinction matters because it separates firsthand experience from speculation. I did not have access to the company's private reasoning, but I stood close enough to watch the sequence unfold and to see what the decision did to the people caught inside it. The experience stripped away another layer of confidence I had once placed in organizational process. Internal investigations were presented as neutral mechanisms for establishing facts, but neutrality requires more than interviews, forms, and approvals. It requires investigators willing to test competing explanations and decision-makers prepared to accept an inconvenient conclusion. A process can appear rigorous while still moving toward a preferred outcome. Documents can be collected, meetings can be held, and procedural boxes can be checked without anyone seriously asking whether the original assumption was wrong. When that happens, process becomes a form of institutional theater rather than a reliable search for truth.

I do not know whether that happened in every part of this case because I was not inside every conversation or decision. From what I could see, however, I had serious doubts that the available evidence had been weighed with the care the consequences demanded. Those doubts did not come from rumor. They came from the footage I viewed, the statement I had heard beforehand, the strong operational record of the business, and the severity of the result. The experience also reshaped how I understood what had happened to me. For years, I had questioned whether I had missed something, mishandled a situation, or simply failed to adapt to an environment everyone else understood. Watching another person pass through the same kind of machinery forced me to consider that the problem might not lie entirely with the people being damaged.

Perhaps the system itself was capable of producing severe consequences without giving those affected a meaningful opportunity to challenge the assumptions driving them. That possibility did not require a coordinated conspiracy or malicious intent from every participant. It required only an imbalance of power, an allegation serious enough to create institutional fear, and a process more comfortable protecting the organization than questioning its own judgment. Once those conditions existed, the machinery could move with very little resistance. Each person could perform a narrow role, follow procedure, and pass the decision forward without anyone taking full responsibility for the final result. The system could destroy a business while allowing everyone involved to believe that someone else had verified the truth. What haunted me most was how quickly a professional identity could disappear inside that machinery. A business built over years could be dismantled in weeks, a reputation could be changed by an allegation before the underlying facts were fully understood, and a family could be destabilized by decisions made in rooms it would never enter.

Returning to Morning Dispatch helped me reconnect with what remained real. The drivers, associates, and frontline managers kept the operation moving despite the failures above them. Their quiet competence mattered more than the slogans displayed on the walls or the leadership principles repeated in meetings. These were people who solved problems without recognition and protected one another without being instructed to do so. They noticed a missing cart before it became a failed route, checked on drivers before departure, and absorbed the pressure created by decisions made far from the floor. Their work reminded me that leadership was not merely the authority to demand results, but the responsibility to protect the people producing them. That part of the operation was still worthy of respect, but the system above it had earned skepticism. The value of the work had never resided in corporate language. It lived in the people who showed up anyway and behaved with professionalism even when the organization failed to do the same.

I left that period with fewer illusions about what Amazon lacked. The company did not lack intelligence, data, technology, surveillance, investigative procedures, or legal resources. What appeared to be missing was the willingness to question decisions once institutional momentum had formed around them. That kind of questioning requires courage: the courage to confront an influential leader, slow down when speed is convenient, and admit that an early judgment may have been wrong. It also requires the courage to distinguish protecting the organization from protecting the truth. Without that courage, procedure becomes a substitute for integrity. The company can point to the existence of an investigation while avoiding the harder question of whether the investigation was fair, complete, and independent of the power surrounding it. Process then protects the institution from criticism rather than protecting people from error.

What I carried forward was not simply bitterness. It was discernment shaped by experience. I had learned what healthy leadership looked like by witnessing its absence, and I had seen how quickly authority could shape a narrative that became difficult to challenge once it hardened into an official conclusion. I had also learned the importance of being precise about what I knew. I could not claim access to every document, conversation, or motive, but I could speak honestly about what I personally saw, what I personally heard, and how those events changed my understanding of the institution around me. What I witnessed was enough to destroy any remaining belief that the system would always correct itself when confronted with conflicting facts.

The experience did not make me weaker. It made me more exact about what I would tolerate, what I would compromise, and what I would never again pretend not to see. Some lessons are learned through success, while others are burned into you by watching power operate without sufficient restraint. This was the latter, and it changed me for good. It left me with no illusions about the difference between a company having a process and a company having the courage to follow the truth wherever it leads.

Finally, a Breath of Fresh Air

After everything that came before, Ted felt like a reset. He joined Amazon as my new manager, coming from UPS with a depth of logistics experience that showed immediately. He understood operations, understood people, and most importantly, understood how those two things were supposed to work together. Once he completed onboarding, I was asked to walk him through how our station actually functioned. I had done these enough times by then that the process was dialed in. We moved quickly, efficiently, without friction. Ted absorbed it all with ease. It was clear early on that he was capable and grounded, not just trained.

What stood out just as much as his competence was his personality. Ted had a sharp sense of humor and a rare ability to laugh at the daily absurdities that came with running a station. He did not take himself too seriously, which made the pressure easier for everyone else to carry. A self-proclaimed coffee fanatic, he showed up one day with a Keurig and more than a hundred flavored K-Cups. His caffeine intake became legend, easily clearing fifteen cups a shift. Somehow, it worked. The energy, the jokes, the constant movement. Ted brought lightness into an environment that had very little of it.

His leadership style was just as refreshing. He was approachable without being soft, supportive without being absent. Associates respected him because he respected them first. The team functioned better. Communication flowed. For a brief stretch, it felt like we had landed in an alternate version of Amazon. One where competence mattered. Where leadership did not rely on fear. Where the work was still hard, but not needlessly punishing. And then, slowly, reality caught up.

About seven months in, Ted started opening up to me about his growing frustration with Amazon's culture. Weekly Business Reviews wore him down the most. The constant criticism. The public finger-pointing. The performative aggression masquerading as accountability. He could not understand why leaders were encouraged to attack each other instead of solve problems together. It ran completely counter to how he had led teams and built results throughout his career.

The breaking point came when Amazon changed his schedule without warning. What had been agreed to as a four-day, ten-hour schedule became five days of eight-hour shifts, weekends included. Saturdays now meant missing his kids' sporting events, something he had been clear from the start he would not sacrifice. On top of that, the promise to place him at a station close to home evaporated. His commute stretched into hours. The concessions kept stacking up, each one small on its own, but collectively impossible to ignore.

When Ted left Amazon, it was not dramatic. It was decisive. He was exactly the kind of leader the company claimed it wanted. Empathetic. Operationally strong. Trusted by his team. And he walked away anyway. Watching him go was disappointing, but I could not fault him for choosing his family and his sanity over a system that refused to bend, even for its best people.

Ted was not pushed out because he failed. He left because the culture made it impossible for leaders like him to stay.

After Ted's departure, my new manager was Darnell, someone I had worked with previously when he ran night sort operations at the station. Back then, we had a solid, professional relationship. I respected his work ethic, his consistency, and the way he handled pressure during overnight operations. Night sort is unforgiving work, and Darnell had earned credibility there. Still, moving from night sort into Dispatch exposed a few knowledge gaps, which was not surprising. Dispatch is its own animal. I worked closely with Darnell to help him bridge those gaps, walking him through the rhythms, decision points, and downstream impacts that only show up once the vans are rolling. To his credit, he absorbed it quickly. He asked the right questions, listened more than he talked, and adjusted without defensiveness.

At first, I was cautious. Darnell had been with Amazon for about four years, working his way up from an Associate into leadership. His trajectory mirrored my own in many ways, which was both reassuring and unsettling. He had also spent time working alongside Mark and Landon, two names that still carried weight for all the wrong reasons. Given what I had lived through, I worried that proximity alone might have shaped his leadership style or normalized behaviors that should never have been acceptable. I braced myself for more of the same, just packaged differently. Those concerns didn't last long.

Early on, Darnell pulled me aside and spoke candidly about what had happened with Mark and Landon. He didn't hedge or soften it. He called the situation what it was, a travesty. He acknowledged that regional leadership had largely sided with them and, in doing so, had effectively shut down any real opportunity for me to advance. Hearing that out loud, from someone inside the system, was validating in a way I didn't realize I needed. Even more surprising was his willingness to openly name the toxic dynamics that leadership preferred to pretend didn't exist. As we continued working together, it became clear that Darnell wasn't just aware of the damage. He actually cared about the people who had absorbed it. That level of empathy was rare in an environment that rewarded detachment.

What truly deepened my respect for Darnell came in a moment that had nothing to do with metrics, reports, or performance reviews. We were scrambling, as usual, trying to pull together a last-minute report for a senior leader who barely understood the operation but demanded answers anyway. In the middle of it, Darnell got a call from his father. He immediately excused himself to take it. The conversation was short, but as he wrapped up, I overheard him say, "Love you, Dad. Can't wait to see you this weekend."

It stopped me cold. As a father myself, with a son, that small exchange hit harder than any leadership principle ever had. It was a glimpse of who Darnell was when Amazon wasn't watching. In a place where people routinely sacrificed family time without question, hearing that simple expression of love and anticipation felt almost subversive. It reminded me that decency still existed inside the building, even if the culture tried its best to grind it down.

For the first time in a long while, I let my guard down because Darnell proved to be an exceptional leader. He was collaborative without being passive, supportive without disappearing, and steady when things inevitably went sideways. We worked long hours through operational problems, staffing gaps, and last-minute escalations, but the work felt shared instead of imposed. I did not just respect him professionally. I genuinely liked him.

As my time at the station wound down and I prepared to transition to another Amazon site, Darnell admitted that he, too, was growing disillusioned. The culture was wearing on him. The politics. The pressure. The constant erosion of values in favor of optics. He told me he was actively looking at opportunities elsewhere. Hearing that was disappointing, but it wasn't surprising. Amazon had a way of pushing out exactly the kind of leaders it claimed to want. People who valued fairness, collaboration, and growth eventually reached a point where staying meant compromising too much of themselves.

After I left, we lost touch. Still, I have no doubt Darnell landed somewhere that recognized what Amazon could not. Someone else undoubtedly benefited from his leadership. Amazon lost yet another thoughtful, principled manager, not because he failed, but because the system failed him. His story reinforced a pattern I had seen too many times by then. The company talked endlessly about leadership, but consistently failed to protect, develop, or retain the people who actually embodied it

Ted and Darnell became bookends to the same truth, seen from two different angles. Ted arrived from the outside, experienced, grounded, and unencumbered by Amazon's internal mythology. He saw the culture clearly and left when it demanded too much of what mattered most to him. Darnell came up from within, shaped by the system but not fully captured by it. He learned how to survive Amazon without losing himself, at least for a while, until the weight of compromise started to press too hard. One chose departure as an act of protection. The other was quietly preparing for it. Different paths, same conclusion.

What their presence gave me, even temporarily, was perspective. After years of distorted feedback, gaslighting, and institutional denial, working alongside leaders who were competent, humane, and self-aware reminded me that I was not imagining the dysfunction. The problem had never been that I could not adapt. It was that I would not contort myself into something unrecognizable just to fit a broken system. Ted and Darnell confirmed that instinct, not through words, but through how they led, how they treated people, and ultimately, how they chose themselves when the culture demanded surrender.

By then, the pattern was undeniable. Amazon did not fail to identify good leaders. It failed to keep them. The environment rewarded endurance over wisdom, compliance over judgment, and optics over outcomes. People who asked hard questions or modeled healthier ways of leading were tolerated only as long as they did not disrupt the narrative. Once they did, the system applied pressure until they either left or learned to look away. Most chose to leave. Those who stayed often paid for it in ways that never showed up on a scorecard.

Looking back, Ted and Darnell represent what Amazon could have been, and what it repeatedly chose not to be. They were not perfect, but they were real. They understood that leadership was not about dominance or performance theater, but about creating conditions where people could do their best work without sacrificing their dignity or their lives outside the building. That kind of leadership should have been protected. Instead, it was treated as expendable.

When I think about my time there now, I do not measure it by promotions earned or titles held. I measure it by moments like those. A manager who walked away rather than miss his kids' lives. Another who said "I love you" to his father without hesitation, even in the middle of chaos. Those moments mattered because they revealed what the culture could not erase completely. Humanity persisted, even when the system tried to grind it down.

In the end, Amazon did not break me. It clarified me. It showed me exactly what kind of leader I am, and what kind I refuse to be. Ted and Darnell were reminders that good leadership still exists, even inside flawed institutions, but it survives there only briefly, like oxygen in a sealed room. Eventually, it has to find its way out to breathe.

The Workforce Amazon Never Saw

Before I left for my new role, the effects of a changing economy were becoming impossible to ignore on the floor. Groceries, rent, gasoline, and other basic expenses had risen sharply, while the pay offered for physically demanding delivery work had not kept pace with the strain of the job. The labor market had also tightened, giving workers more reason to reject positions that required ten-hour days, constant lifting, and relentless time pressure without providing enough financial stability in return. The DSP program felt that shift immediately as driver turnover accelerated and recruiting replacements became increasingly difficult.

Amazon responded by increasing hiring events and pressing stations and DSPs to fill vacancies quickly. But the emphasis on speed often came at the expense of finding applicants who fully understood what the job required.

During interviews, I met candidates who had not been given a realistic picture of the work. Some were unprepared for the physical demands. Others did not appear comfortable with the communication, navigation, and independent decision-making the role required. In several cases, it was clear that recruiters had focused on moving people through the hiring process without adequately considering whether the position was safe, sustainable, or appropriate for the individual.

The problem was not that applicants were unwilling to work. It was that Amazon was trying to fill an increasingly difficult job without addressing the conditions that made people leave it.

This was not a judgment of the people themselves. It was an indictment of a system so desperate to fill seats that it stopped being honest about what the job actually required.

DSPs were left in an impossible position. They needed drivers desperately, but hiring people who could not safely or reliably perform the work only created downstream failures.

Many applicants were turned away, not out of cruelty, but out of necessity. Routes still had to go out. Customers still expected deliveries. And Amazon still demanded full coverage.

As staffing shortages deepened, the financial pressure on DSPs intensified. With fewer drivers, revenue dropped. At the same time, operating costs continued to rise. Insurance, fuel, vehicle maintenance, and compliance expenses all climbed. The math stopped working. Yet Amazon's expectations did not adjust. DSPs were still held to the same route commitments, the same performance metrics, the same zero tolerance standards, regardless of the reality on the ground. That disconnect pushed DSPs toward survival mode.

One workaround became increasingly common. Drivers would load multiple routes at the station, then stage vans at a central location within the delivery area. Once a driver completed their assigned route, they would return to that staging point and pick up additional packages from another van. On paper, it helped maintain route completion. In reality, it shifted the burden entirely onto the drivers. They worked longer hours. Took fewer breaks. Drove while exhausted. The physical and mental toll added up quickly.

The consequences were predictable. Safety incidents increased. Fatigue related mistakes became more frequent. Customer complaints about late or misdelivered packages spiked. Everyone in the system felt the strain. DSP owners were fighting to stay solvent. Drivers were burning out at alarming rates. Stations were scrambling daily to plug holes that should never have existed in the first place.

The system was not breaking because people stopped caring. It was breaking because it was no longer designed to support the people holding it together.

I felt compelled to act, but I had no idea where to start. The problem felt massive and tangled, stretching across Amazon's hiring practices, route management decisions, and the growing economic pressure crushing DSPs and drivers alike. Every lever I could see was connected to something else. Fix one thing and two others seemed to break. Change was clearly necessary, but the scale of it all left me frozen. I couldn't shake the question of whether Amazon's systems or its leadership would even allow meaningful reform, even if someone managed to design it.

The answer, or at least the spark, came from the last place I expected.

One afternoon, while shopping at Trader Joe's, I was searching for a specific item when I noticed a Crew Member stocking shelves nearby. What caught my attention immediately was his demeanor. He was smiling. Not the forced, customer-service smile, but a genuine one. He moved with purpose, clearly engaged in his work, radiating a sense of pride in what could easily be a repetitive, thankless task.

I approached him from behind and asked where I could find the item I was looking for. No response. I asked again, louder this time. Still nothing. Annoyance crept in. My first thought was, how rude. Determined to get his attention, I stepped in front of him and repeated the question, this time with a sharp edge I did not bother to hide.

He looked at me, smiled warmly, and pointed to his ears. That was the moment it clicked: he was Deaf. Embarrassment hit instantly. Without thinking, I repeated the question more slowly, as if changing my pace would somehow fix my ignorance. He simply smiled, motioned for me to follow, and led me directly to the item I needed. Once I had it, he gave me a thumbs-up and returned to work, still smiling, still focused, and completely unfazed by the interaction.

I walked away feeling like a complete jerk, but I could not stop thinking about what I had just seen. A few moments later, I doubled back and watched him more closely. Another Crew Member joined him and began communicating in Sign Language. They worked seamlessly together, exchanging quick signs, grabbing boxes, stocking shelves, moving efficiently without friction or frustration. The Deaf Crew Member never lost his energy. Never lost his smile. Never slowed down.

That was when it fully landed. Here was a demographic routinely overlooked, sidelined, or quietly excluded from the workforce, not because of a lack of ability, but because of society's unwillingness to adapt. Watching him work, I saw no barrier to productivity and no limitation on effectiveness. If anything, he seemed more engaged and fulfilled than many employees I had seen in similar roles elsewhere. His Deafness was not an obstacle. It was simply one aspect of who he was.

And then the question formed clearly: Why couldn't this work at Amazon? Amazon had always struggled with workforce stability, especially in last-mile delivery and station roles. The work was physical, the pace relentless, and the turnover brutal. Yet here was an untapped pool of capable, motivated people being excluded, not because they could not do the job, but because companies refused to meet them halfway on communication and inclusion. For the first time in a long while, I saw a path forward that was not rooted in squeezing people harder, but in widening the door. That realization stayed with me.

By the next day, I knew I could not ignore what I had seen. I reached out to the Recruiting Group Lead and proposed a targeted approach to driver-hiring events focused specifically on the Deaf community. I explained the potential plainly. Here was a group of capable people who were routinely overlooked by traditional recruiting, yet I had just seen firsthand how effectively someone could perform when the environment made even a modest effort to accommodate communication differences. It seemed like an opportunity to address several problems at once: chronic driver shortages, high turnover, and the constant struggle to find people willing and able to sustain the demands of last-mile delivery.

Her response was immediate and dismissive. She called the idea dumb and said it would never work at Amazon. There was no real discussion, no curiosity, and no attempt to understand what I had observed. The idea was rejected before it was tested.

I remember thinking, what a dumbshit attitude, especially given how badly her recruiting efforts had already failed. We were struggling to fill roles, burning through drivers, and constantly searching for new ways to stabilize staffing. Yet when a different approach appeared, the instinct was not to explore it, but to shut it down. That reaction frustrated me more than the rejection itself. Amazon talked constantly about innovation, inclusion, and thinking differently, but in that moment, none of those principles seemed to matter. The safest response was still to defend the existing approach, even when the existing approach was not working.

When I escalated the idea to upper leadership, the reaction was no different. They dismissed it almost immediately and told me it was a waste of time, that it would never work, and that I needed to stay in my lane, focus on Dispatch, and stop chasing what they considered worthless initiatives. There was no discussion about testing the idea, no interest in defining what success might look like, and no curiosity about what I had observed. The decision had effectively been made before the conversation began.

What frustrated me most was how familiar the response felt. I had seen the same pattern repeatedly: a process fails, someone closest to the work proposes a different approach, and leadership treats the challenge to the existing system as the problem. Curiosity disappears, unconventional thinking becomes a liability, and the people with firsthand knowledge are told to execute rather than question. Even when the existing approach is clearly falling short, defending it can feel safer than admitting another solution might work.

This time, something in me snapped into clarity. Fuck it. If they were not willing to support the idea, I would find a way to pursue it myself. I knew the staffing problems we were facing, how difficult it was to find dependable drivers, and how quickly turnover could destabilize an operation. I had also seen firsthand that Deafness was not the barrier. The real barriers were communication, access, and an unwillingness to adapt.

Their refusal to explore the idea did not convince me it was wrong. It convinced me they were unwilling to see an opportunity that did not fit their existing assumptions. I stopped asking for permission and started thinking about how I could prove the concept on my own. If the organization would not test the idea willingly, I would find a way to give them results they could not easily dismiss.

Selling the idea to the DSP owners was never going to be easy. From their perspective, the risks were obvious. Liability concerns. Communication barriers. Fear of the unknown layered on top of an already fragile business model. Still, relationships matter, and I had spent years building trust with them in the trenches. I was not pitching a theory or a corporate initiative. I was asking for a pilot. A few owners agreed to give it a shot, not because Amazon told them to, but because they trusted me and were desperate for solutions that actually worked.

Once I had tentative buy-in from the DSPs, the next challenge was finding candidates from the Deaf community. That part came together through pure luck. A former colleague of mine was now running Amazon's ASL program for Sort Associates. He had already been experimenting with integrating Deaf Associates into the sortation line and was facing many of the same barriers I had run into. Leadership talked about inclusion, but offered very little support when it came time to actually make it work. We connected quickly. He understood the resistance. More importantly, he believed the effort mattered.

He became an invaluable resource. Through him, I was introduced to organizations outside of Amazon that were deeply connected to the Deaf community and actively focused on employment access. For the first time since the idea had taken shape, it felt like there was a real path forward.

I reached out directly to the Employment Development Department, the Orange County Deaf Equal Access Foundation, the Tri-County Greater Los Angeles Agency on Deafness, and the Center on Deafness Inland Empire. I explained what I was trying to build in straightforward terms: the work, the expectations, and the support I was prepared to put in place. Their response was immediate and overwhelmingly positive. Each organization expressed genuine enthusiasm for an initiative that treated Deaf candidates as capable contributors rather than accommodations to be managed. That reaction reinforced what I already believed: the idea had merit, regardless of Amazon's skepticism, and there were organizations ready to help make it real.

From there, the effort gained momentum quickly. I needed recruiting materials that actually spoke to the community I was trying to reach, so my colleague running the ASL program and his team created a clear, thoughtful ASL video explaining the hiring event, the job itself, and what candidates could realistically expect. There was no corporate gloss or vague recruiting language. It was direct, accessible information delivered with respect, giving candidates a clear understanding of the opportunity before they ever walked through the door.

At the same time, I leaned on my Driver Training Team. This was a group of exceptionally capable people who consistently delivered results while receiving little recognition from leadership. Within days, they developed training materials tailored specifically for Deaf drivers. They approached it the same way they handled every challenge. They focused on what would work and made it happen.

They also helped identify practical tools that would make the effort viable. Sorenson provided professional ASL interpreters. Purple Communications, through its P3 app, offered real-time ASL support via Video Relay Service on smartphones. These were not experimental solutions. They already existed. All Amazon had to do was allow their use.

With the materials complete, a registration email drafted, and a hiring date locked in, I sent everything out to the organizations I had contacted. Within days, my inbox started filling up. Interested candidates. Thoughtful questions. Requests for next steps. Momentum was building.

I kept the DSP owners informed, walked them through how P3 worked, coordinated ASL interpreter coverage through Sorenson, and made sure everyone understood exactly how the hiring event would run. On paper, everything was in place. In reality, I had no idea how it would unfold, and all I could do was wait. For the first time in a long while, I felt something close to cautious optimism.

The hiring event exceeded every expectation I had. Twelve Deaf candidates showed up, and every one of them walked away with a job offer from the DSP owners. On paper, it was a success, and operationally, the concept worked. But one moment from that day eclipsed every metric and result and has stayed with me ever since.

One of the candidates stood out immediately. A young man who walked into the station carrying himself like someone who had already been told no too many times. His shoulders were hunched forward, his posture closed in on itself, as if he were trying to take up as little space as possible. His eyes rarely lifted from the floor. There was a heaviness about him that went far beyond nerves. It looked like exhaustion. The kind that comes not from a bad week, but from a life spent pushing against barriers that never seem to move.

I remember thinking this was not just about unemployment. It was about what constant rejection does to a person, how it chips away at confidence and slowly convinces you that effort is pointless and hope is dangerous. I could not know the full scope of what he had endured, but I did not need to. His body language told the story. This was someone who had learned to expect disappointment and had armored himself accordingly, yet he still went through the interview.

When the DSP owner made the decision and offered him the Delivery Associate role, the transformation was immediate and impossible to miss. His face changed first. Then his posture. Then his entire presence. The weight he had been carrying seemed to fall away all at once. What replaced it was pure, unfiltered joy. The kind that catches in your throat when you see it because it is so raw and so rare.

He looked at me, eyes filled with tears, and walked over without hesitation. He wrapped his arms around me and held on. No words. No explanation. Just gratitude, relief, and disbelief all tangled together. In that moment, language was irrelevant. Everything he needed to say was already there.

Even now, writing this, I feel it again. Watching someone move from visible despair to genuine happiness in the span of a few minutes is something that stays with you. That offer was not just a job. It was validation. It was dignity. It was proof that he was seen and valued in a world that had repeatedly told him otherwise.

That was when I knew this initiative mattered far beyond metrics or headcount. It was never just about filling routes or solving a staffing problem. It was about restoring something far more fragile. Hope. For people who had been overlooked, underestimated, or dismissed for reasons that had nothing to do with their ability to work hard or contribute.

That moment reminded me why I refused to let the idea die when leadership dismissed it, and it reinforced a truth Amazon often seemed unable to grasp. Inclusion is not a slogan or a line in a corporate presentation. When it is taken seriously and built into the way people are hired, supported, and trusted, it can change lives.

The real test, of course, came once the vans rolled out. Everything before that had been preparation, planning, and belief. The road would tell us whether the idea actually worked, and it did not take long to get the answer.

The Deaf drivers performed exceptionally, in ways that were impossible to ignore and even harder to dismiss. Not a single Deaf Delivery Associate returned with undelivered packages. Routes were completed cleanly and on time. Many finished early and immediately asked Dispatch if they could help others who were behind. There was no hesitation, no prompting, no sense of doing the bare minimum. They treated the work with seriousness and pride, and the results showed up in the metrics day after day.

What stood out just as much as their efficiency was how they carried themselves. Like the Crew Member I had watched at Trader Joe's, they worked with genuine joy. Not forced positivity. Not performative engagement. Real joy. The kind that comes from finally being given a fair shot and being trusted to do the job. It changed the tone on the floor. Their energy lifted the people around them, and other drivers took notice. Respect followed naturally, not because of a policy, but because performance earns it.

During Dispatch, those moments stayed with me the most. As they headed out for their routes, several of the Deaf drivers would pause, look back toward me, and sign love by crossing their arms over their chest. Others used the iconic ILY sign, thumb, index finger, and pinky raised, smiling as they held it up. No words were exchanged. None were needed. It was gratitude, connection, and trust expressed in the simplest and most human way possible.

Over time, I held nine additional Deaf hiring events. By the end of it, more than fifty Deaf Delivery Associates had been hired and successfully placed with DSPs. They were not treated as exceptions. They were not handled with kid gloves. They were held to the same standards as everyone else, and they met them. Consistently.

Out of everything I accomplished during my time at Amazon, this stands at the very top. Not because it looked good on paper, but because it mattered. I felt genuine pride knowing I had helped open a door for people who had been overlooked for far too long, not due to lack of ability, but because the system was unwilling to adapt. The Deaf community brought more than reliability and work ethic. They brought heart. They brought joy. And they reminded everyone around them what meaningful inclusion actually looks like when it is done right.

After I left the station, the program ended. Without leadership support, recruiting buy-in, or anyone willing to fight for it the way I had, the initiative quietly disappeared. It was abandoned not because it failed, but because Amazon never truly valued it enough to sustain it. The decision became another example of how the company could recognize an idea's value in practice yet still fail to support the people and programs that made the organization better.

Still, that does not take away what was accomplished. The smiles. The pride. The confidence I watched grow. The success of those fifty individuals will always stay with me. That chapter of my career was proof that change was possible, even inside a system that resisted it at every turn. And it remains one of the proudest things I will ever carry with me.

By the time that chapter closed, the lesson was impossible to ignore. The system did not fail because it lacked data, technology, or scale. It failed because it lacked the willingness to listen when solutions came from the margins instead of the org chart. What the Deaf driver initiative proved was not just that inclusion was possible, but that it was operationally stronger. The work was done better. The culture was healthier. The results spoke for themselves. And still, the moment it required sustained leadership courage, the effort was allowed to die.

That contradiction stayed with me. Amazon loved outcomes, but only when they arrived through approved channels. Innovation was celebrated in speeches and leadership principles, but punished when it disrupted comfort, hierarchy, or control. The fact that more than fifty people found stability, dignity, and pride through that work mattered far more to me than any metric ever could. Those drivers did not need slogans or panels or press releases. They needed a door opened and the chance to prove themselves. They did exactly that.

Watching the program disappear after I left was disappointing, but not surprising. By then, I understood the pattern. Change was tolerated as long as it did not challenge authority. Once it did, it became inconvenient. But even knowing how it ended, I would do it again without hesitation. Some work is worth doing even when you know the system will not protect it.

That chapter marked a turning point for me. It clarified what kind of leader I was willing to be and what kind of environment I could no longer pretend was enough. The success of those drivers did not just restore hope for them. It restored something in me. A reminder that leadership, when stripped down to its core, is not about permission or recognition. It is about seeing people clearly, acting with conviction, and standing by the results, even when the system refuses to.

Efficiency Has a Cost

As my career at Amazon progressed, I moved through roles that steadily pulled me farther from the floor and closer to the architecture of the operation itself. What began as hands-on, day-to-day problem solving gradually turned into something broader. I was no longer just reacting to breakdowns. I was learning how and why they happened. Eventually, I landed in a role that tied everything together, one that gave me a full view of how individual decisions on the floor cascaded through the network. It was in that final stage that I came to understand the true weight of one of Amazon's most critical commitments: the Promised Delivery Date (PDD).

At first, PDD looked like just another metric, another number leadership tracked obsessively. Over time, I realized it was the connective tissue holding the entire operation together. Every associate, manager, engineer, and executive is ultimately working toward that single promise. Meeting it was not just about speed or efficiency. It was about trust. When Amazon failed to hit PDD, teams mobilized immediately, tearing apart processes and data to understand exactly where the system had broken down. What struck me most was how small errors could create outsized consequences. A misloaded package, a missed scan, a brief delay in sortation could ripple outward, disrupting routes, compounding delays, and undermining the customer experience Amazon had built its reputation on.

One of the most persistent threats to PDD was missorts, packages sent to the wrong downstream location within the network. When I stepped into my final role as a Program Manager, I was tasked with leading efforts to reduce them. On the surface, a missort seemed trivial. In reality, it was anything but. Each one carried an average cost of more than three dollars, and at Amazon's scale, those errors translated into hundreds of millions of dollars in losses every year. Beyond the financial impact, missorts consumed capacity, strained logistics, and eroded reliability in ways customers never saw but absolutely felt.

The mission was straightforward in theory. Identify patterns, uncover root causes, and provide actionable guidance to the North American network. In practice, it meant applying basic operational principles at enormous scale. Large data sets and machine learning models were used to flag vulnerabilities before errors occurred. The real challenge was turning analytics into action. Insights only mattered if they could be translated into changes that associates could realistically execute in the middle of a relentless workday. That required close coordination with fulfillment centers, sortation sites, and regional teams to ensure solutions worked on the floor, not just on a dashboard.

One of the hardest balances to strike was speed versus accuracy. Associates were under constant pressure to move faster, and that pressure inevitably produced mistakes. Leadership often defaulted to demanding improvement without addressing the conditions that made errors unavoidable. Telling people to be more careful was meaningless if the system itself left no margin for care. Real improvement required restructuring workflows, refining training, and adjusting processes so accuracy was achievable under real operating conditions.

As the analysis deepened, it became clear that missorts rarely had a single cause. Some failures were technical, unclear labels, software issues, scanner misreads. Others were human, fatigue, burnout, and the cumulative effect of nonstop productivity demands. Each finding reinforced the same truth. The efficiency that powered Amazon's success also created the conditions for failure. Fixing the problem meant improving precision without increasing strain on the people doing the work.

The impact of those efforts was real. Missorts dropped. Costs came down. Delivery reliability improved. Customer trust was reinforced. Just as important, the changes reduced unnecessary friction for associates. Tasks became more manageable. Pain points that had quietly made the job harder were removed. The work was still demanding, but it was less punishing. In an environment like Amazon, even small improvements made a measurable difference.

Looking back, the arc from frontline associate to Program Manager taught me more than logistics or data analysis. It revealed how fragile the balance between efficiency and humanity really was. Amazon's operation was extraordinary in its complexity and scale, capable of moving millions of packages with astonishing precision. But beneath all the automation and optimization, it remained a system powered by people. People working under immense pressure, whose contributions were largely invisible once the box landed on a doorstep.

Seeing that system from the inside gave me a deep respect for the workers who kept it running. The scale was extraordinary, but so was the amount of human effort required to make that scale possible. Behind every route, metric, scan, and delivery promise were people adapting in real time, solving problems the systems did not anticipate, and absorbing the consequences when plans fell short.

It also reinforced a belief that had followed me throughout my career: efficiency should never require sacrificing humanity. Strong systems should make people better at their work, not reduce them to inputs measured only by speed, output, and compliance. Technology can remove friction, standardize processes, and improve performance, but those gains lose their meaning when they come at the expense of judgment, dignity, or basic human limits.

At Amazon, that balance was often lost. The company excelled at building systems that moved faster, measured more, and demanded constant improvement. What it struggled to build were systems that protected the people expected to sustain that pace. The machinery of efficiency kept advancing, while the human infrastructure behind it was too often left to absorb the strain.

As I moved forward, I carried those lessons with me. Amazon gave me an unparalleled education in operations, scale, and problem solving. It also left me with a clear conviction. True operational excellence is not just about speed, cost, or optimization. It is about designing systems that work for the people who power them. That belief continues to guide how I approach leadership, long after I left the building.

By the end of that chapter of my career, I no longer viewed Promised Delivery Date as just a performance commitment or a customer-facing metric. PDD had become a kind of truth serum for the organization. It exposed where systems were resilient and where they were brittle. It showed exactly how much slack Amazon was willing to remove from its operation and who ultimately paid the price for that efficiency. Every missort, every rushed scan, every exhausted associate was a signal, not of individual failure, but of a system operating at the edge of human tolerance.

The work to reduce missorts proved that Amazon was capable of real, meaningful improvement when it chose to be. Costs came down. Reliability improved. Life on the floor became marginally more manageable. But those gains only happened because the problem was approached honestly, with a willingness to adjust processes instead of simply demanding more effort. That contrast stayed with me. The company knew how to fix problems without breaking people. It just did not always choose to do so.

What unsettled me most was how invisible that tension remained once the package reached the customer's doorstep. From the outside, the system looked flawless. Fast. Precise. Effortless. From the inside, it was clear how much pressure had been compressed into every second of movement. The people absorbing that pressure rarely appeared in the narrative of success. Their exhaustion, judgment calls, and quiet problem-solving vanished the moment the delivery was completed.

That perspective fundamentally changed how I understood leadership. Efficiency, on its own, is not an achievement. It is a tradeoff. Every optimization decision asks an implicit question about where the burden will land. My time moving from the floor to the architecture of the network taught me that the most dangerous failures are not the ones that show up in metrics. They are the ones that slowly normalize strain, burnout, and silence in the people holding the system together.

I left that role with a deeper respect for the complexity Amazon had built and an equally clear-eyed understanding of its limits. Scale does not absolve responsibility. Data does not replace judgment. And no system, no matter how advanced, can outrun the human cost it creates indefinitely. That realization became one of the most important lessons I carried forward, shaping not just how I led, but how I decided where my time, energy, and integrity were worth investing next.

The Role That Explained Everything

My final role at Amazon pulled together everything I had learned over the years and placed me at the center of the system I had once worked inside from the floor. I was responsible for establishing and owning a North American Network Operations Center Quality function, defining how quality would be measured, governed, and escalated across sort centers and downstream delivery networks. For the first time, I had the vantage point to see how decisions made in one corner of the network rippled outward, affecting delivery performance, cost, and customer trust at scale.

I owned the network-wide quality strategy for North American sort centers, designing the analytics, KPIs, and escalation mechanisms used to identify breakdowns before they reached the customer. That work required constant collaboration with software engineering teams, senior leaders, transportation partners, planners, and program managers across multiple organizations, all focused on improving delivery estimate accuracy and reducing downstream defects. It was less about chasing metrics and more about understanding where the system was quietly failing and why.

A large part of the role involved leading cross-functional planning efforts that spanned transportation, technology, and operations. Volume forecasting, capacity planning, and labor efficiency were no longer abstract concepts. They were daily levers that determined whether the network could absorb demand without breaking the people inside it. I was also responsible for communicating upcoming network changes, programs, and events that would impact short-, medium-, and long-term operations, ensuring leaders across the network were aligned before disruptions occurred.

I led and managed a team of Process Assistants tasked with executing standard work and supporting new initiatives as they were introduced. Together, we owned network-level performance management, building executive dashboards, decision frameworks, and analytical tools that senior leadership relied on to steer quality, cost, and delivery outcomes across North America. When failures occurred, my team and I conducted deep dives into process breakdowns, not to assign blame, but to identify sustainable fixes that could be scaled without adding unnecessary burden to frontline teams.

Data quality was foundational to everything we did. I worked closely with data engineering teams to expand data collection, refine existing sources, and improve accuracy so insights reflected reality on the ground rather than what leadership wanted to see. At the same time, I focused heavily on mentoring and feedback, trying to foster a culture of continuous improvement that valued learning over optics.

The role also required close partnership with internal oversight groups responsible for sort centers, USPS, and third-party logistics carriers. Aligning those relationships was critical to maintaining consistency across a network that depended on thousands of interconnected handoffs every day.

From the outside, it was an impressive position. From the inside, it was the clearest view I ever had of both Amazon's brilliance and its blind spots. The systems were powerful. The scale was staggering. But even at this level, it was impossible to ignore how often efficiency was prioritized over sustainability, and how frequently the human cost of keeping the machine running was treated as secondary. That final role did not just cap my career at Amazon. It crystallized everything I had come to understand about the company, its culture, and the tradeoffs it was willing to make.

That role should have felt like arrival. Instead, it felt like resolution. Standing at the center of the network, with visibility few ever get, confirmed what years on the floor had already taught me in fragments. Amazon is extraordinary at building systems that scale, detect failure, and correct course at breathtaking speed. It is far less disciplined about asking what those corrections cost the people absorbing them. The higher I went, the clearer that tension became. My final role did not leave me disillusioned about operations. It left me convinced that operational excellence without human sustainability is a temporary victory at best. What I carried with me when I stepped away was not just technical knowledge or strategic insight, but a sharpened understanding of where systems succeed, where they fail, and why leadership matters most precisely when the numbers look good.

Inherited Failure

Stepping into the Program Manager role to develop a Team tasked with identifying metric misses across every sort center in North America did not feel like a promotion. It felt like being dropped into the middle of a mess leadership had ignored for far too long. The team I inherited consisted of twenty Process Assistants (PA) assigned to support another team by handling SIM trouble tickets, all well compensated, yet almost none showing any real interest in doing the work they had been responsible for long before I arrived. What I walked into was not inexperience or normal growing pains, but entitlement, a complete absence of accountability, and a culture that had learned it could operate without consequences and had been doing so comfortably for quite some time.

That reality became impossible to ignore on my first day. I walked into the office expecting to see people working and instead found desks converted into beds, blankets and pillows laid out, team members asleep in the middle of the workday. No laptops were open. No analysis was running. The people responsible for identifying metric failures across North America were producing nothing, reviewing nothing, and flagging nothing. There was no sense of urgency and no sense of shame. My first reaction was disbelief. Aren't they supposed to be working? That disbelief quickly turned into concern when it became clear this was not an isolated lapse, but routine behavior within a team whose inaction directly translated into blind spots, bad data, and missed signals leadership relied on to make decisions at scale.

The question that stayed with me was not just what I had walked into, but how leadership had allowed it to exist at all. That became apparent almost immediately when several managers from other teams, unprompted, told me "Good Luck" with this group and noted they had been a problem for some time. It was not encouragement. It was a warning. As I began managing the team day to day, the depth of the dysfunction became impossible to ignore. Chronic tardiness was treated as optional, with people drifting into shifts and meetings as if schedules were suggestions rather than requirements. Some manipulated their hours, clocking in early or staying clocked in while doing nothing, while others disappeared entirely without notice, accumulating unauthorized absences far beyond what any functional organization would tolerate. Sleeping on the job was not an exception or an anomaly. It was baked into the culture.

Distraction dominated the environment. Phones were always out, and what really pissed me off was that almost immediately, several leaders stopped by my desk to inform me that someone was using a cell phone. WTF? Where was this concern before I arrived? Why had this behavior been tolerated long enough to become routine, only to suddenly matter once it was no longer someone else's problem? Browsers were open to anything but work, and productivity consistently came second to convenience. The office reflected that mindset. Food was left out, personal items were scattered everywhere, and the space looked less like a professional workplace and more like a neglected dorm room. This was the team responsible for quality, operating without standards, structure, or oversight, and doing so with a level of comfort that suggested they were proud of how little was expected of them.

What stood out most was not laziness by itself, but how normalized it had become. This behavior did not appear overnight. It existed because no one above them had been willing to confront it. Expectations had either never been clearly set or had been set and quietly ignored, and when they were violated, nothing happened. Leadership had not failed accidentally. It had failed by looking the other way, again and again, until avoidance became policy.

And now the mess was mine to clean up. When I raised these concerns, leadership responded with a single hollow directive: "You need to set up guardrails." Guardrails. The word landed with a thud. There was no clarification, no acknowledgment of how long these issues had existed, and no explanation for why those guardrails had never been put in place before I arrived. What leadership was really saying was clear. The dysfunction was now my problem, and whatever fallout came from addressing it would be mine as well.

It became obvious very quickly that upper management had already given up on this team. Rather than confront the behavior directly, reset expectations, or absorb the conflict themselves, they passed the responsibility downward without authority, cover, or support. I was not stepping into a leadership role so much as inheriting abandonment. Instead of being empowered to lead, I was expected to babysit a group that had been allowed to operate without structure, accountability, or consequences for years. Standards had eroded so badly that enforcing even the most basic expectations felt controversial. This was not an opportunity to build something new. It was an assignment to quietly contain a problem leadership had knowingly allowed to fester.

The entitlement surfaced immediately. During our first full crossover meeting, with both front half and back half shifts present, I began laying out clear expectations around attendance, workload, and professionalism. I did not even finish the sentence before one PA interrupted me and said, without hesitation, "I'll do the bare minimum, and if you expect more, I'll complain to HR." There was no embarrassment and no attempt to soften the message. It was a threat, delivered plainly and confidently, and it set the tone for everything that followed. What should have been a fifteen-minute alignment meeting disintegrated into a two-hour airing of grievances, with each person piling on complaints about a lack of direction, a lack of recognition, and managers who would arrive, disappear into their offices, shut the door, and treat the team as an inconvenience rather than people.

But the real centerpiece of the meeting was not process or workload. It was the previous manager, now my current manager, and the open contempt this team had for that leadership role. The hatred was raw, unfiltered, and relentless.

It surfaced casually, repeatedly, and without hesitation, as if trashing prior leadership had become a shared language inside the group. Sarcasm replaced professionalism. Disrespect was normalized. Every failure was traced back to that role personally, every grievance sharpened into an indictment of competence and credibility. This was not frustration. It was scorched earth resentment, cultivated over time and spoken aloud with zero fear of consequence.

Standing there, it became clear I was not inheriting a disengaged team or even a dysfunctional one. I was stepping into a culture where leadership credibility had been obliterated, where authority was openly mocked, and where resentment had hardened into identity. Whatever this team had been allowed to become had already crossed a line, and the damage was no longer abstract. It was alive in the room. What looked on paper like an opportunity felt, in real time, like walking into a hostile environment that had been decaying for years, and I knew instinctively that this level of bitterness does not stay contained. It leaks. And eventually, it looks for a new target.

The moment expectations rose to anything resembling a normal workload, HR complaints followed. Claims of being "overworked" appeared immediately from several team members, led by the same PA who had openly stated they would do the bare minimum, despite peers handling identical volume without issue. Performance conversations were reframed as personal attacks. Accountability was labeled hostility. The message was unmistakable. If you resisted standards loudly enough, the system would intervene on your behalf. Effort became optional because escalation worked.

When the HR manager finally engaged me, it was not to ask about the documented patterns of abuse, time manipulation, or chronic nonperformance, all of which were already visible and trackable. Instead, I was told to lower workload expectations and offer more flexibility. You have got to be "Fucking Kidding" me. HR did not question the complaints. They did not examine the data. They did not ask how a Team could function without enforcing basic accountability. They simply reinforced the behavior that had broken the team in the first place. In that moment, it became clear I was on my own. The system did not protect standards or outcomes. It protected noise, rewarded resistance, and punished anyone who tried to restore order.

What I was witnessing had nothing to do with individual personalities. It was the predictable outcome of a culture that had trained people to weaponize policy in place of performance. Leadership's long-standing refusal to draw boundaries had created an environment where compliance was negotiable, effort was optional, and resistance carried no cost. By then, the truth was no longer debatable. This was not a broken team in isolation. It was a system functioning exactly as designed, enabling the behavior, shielding it from correction, and then transferring the consequences downstream to whoever was willing, or unlucky enough, to take ownership.

This Is What Leadership Looked Like

What happened next confirmed exactly where leadership's priorities actually lay. HR did not step in to address open insubordination or the behavior that had already hollowed out the team. Instead, I was told to lighten the workload. The message was unmistakable. Push back, and the system would push back harder. Effort was negotiable. Accountability was optional. Resentment followed almost immediately. The PAs who showed up and did the work were forced to carry more, while those who resisted were rewarded with lower expectations. Morale, already fragile, eroded further under a system that punished effort and protected defiance.

By then, it was clear leadership had no intention of offering real support. I had walked into a problem that had been building for a long time, and instead of being given the authority to fix it, I was expected to manage around it, absorb it, and normalize it. I was no longer surprised by that reality. I was just tired. What really pissed me off came next, when managers who had watched this team unravel long before I arrived suddenly started coming to me with questions. Why is this PA on their phone? Why did I see this PA at Starbucks during their shift? The implication was absurd. As if I had created the culture overnight. As if this dysfunction had not been visible to everyone for years. Once there was a name and a face attached to the mess, the blame finally had somewhere to land, and of course it landed on me.

That same pattern surfaced elsewhere. The team constantly complained about a toxic culture they attributed to prior leadership, blaming a previous manager they openly referred to as Cruella. They described a pattern of neglect, constantly shifting expectations, and deliberate avoidance, unloading each grievance in detail as if they had been waiting for someone to finally listen. The unspoken message was always the same. None of this was their fault. At first, I assumed this manager was long gone. That assumption did not last. Cruella was still there. Cruella was my direct manager. That realization explained far more than I wanted it to. The chaos the team described was not a closed chapter. It was the foundation I was now standing on, and it had never been corrected or contained.

At first, that legacy showed up as absence. Expectations without follow-through. Damage without ownership. But it did not stay passive for long. As I began building real structure and momentum, the influence shifted from neglect to control. Cruella's style revealed itself through obsessive micromanagement and a fixation on details that did not matter. Real work was constantly derailed by demands for exhaustive explanations tied to discrepancies rooted in system flaws well outside our control. Emails arrived in all caps, frantic and punitive. How are you pulling that data? My numbers do not match. Everything must be reconciled by Monday end of day.

Defining the process meant working closely with teams across the Amazon network, partnering with operations, tech, and analytics to align on how defects would be identified, how ownership would be assigned, and how communication back to each SC would drive real correction instead of noise. Drawing on my background in data modeling, I built models that surfaced metric defects clearly and tracked resolution end to end across all SCs, not just whether a miss occurred, but whether it was acknowledged, owned, and actually fixed. For the first time since stepping into the role, I felt intellectually engaged. I was proud of what I was building, and for a brief stretch, the work made sense.

Once the process was defined, I translated it into clear Standards of Work for the PAs. The goal was independence, not oversight. I wanted them to have a roadmap that showed exactly how to deep dive the data, identify defects, validate root cause, and track resolution without hand holding. In that work, I felt like myself again. Focused, capable, and effective. For a brief window, the work made sense, and doing it well brought real satisfaction, even as the larger environment continued to unravel around it.

That sense of enjoyment faded quickly under my manager's influence. The management style I experienced was defined by obsessive micromanagement and a fixation on details that did not matter. Meaningful operational work was repeatedly interrupted by demands to explain discrepancies caused by system failures beyond our control. Emails arrived in all caps, frantic and punitive. How are you pulling that data? My numbers do not match. Everything must be reconciled by Monday end of day. Nothing was ever truly resolved. Each supposed fix was followed by another fire drill, another forced correction, another deadline designed to assert control rather than improve outcomes.

That behavior did not stop at the office. Texts came late at night and on weekends. Demands for reports first thing Monday morning arrived as if my time belonged to someone else by default. There were no boundaries, no off hours, and no relief from the pressure. Over time, it stopped feeling accidental. It began to feel intentional, as if the goal was not to support the work, but to grind me down until failure became inevitable.

It was not just exhausting. It was corrosive. I was worn down and emotionally wrecked by the constant pressure from both Cruella and the team I was expected to manage, all while operating without any meaningful support from above. Every time I raised a concern, it disappeared into silence. No follow-up. No engagement. No acknowledgment. Over time, that silence became its own message.

The weight of it all pushed me into a mental space I never expected to reach. I began having thoughts I had never entertained before, moments where the idea of simply not having to carry it anymore felt like relief. The pressure, the isolation, and the sense of being trapped without support or an exit became overwhelming. No one intervened. The behavior driving it was never questioned. It was protected. I was not failing. I was drowning.

Cruella's legacy was fear and frustration, but over time a deeper truth became impossible to ignore. This was never just about one toxic manager or one broken dynamic. It was about a culture that tolerated this behavior, enabled it to persist, and quietly rewarded it by refusing to intervene. The dysfunction did not survive in spite of the system. It survived because the system made room for it.

When I stepped into the role, I was promised support. That promise sounded reassuring on paper and convincing in conversation, but it collapsed the moment support was actually required. The second accountability became uncomfortable, the second pressure moved upward instead of downward, those assurances vanished. What remained was silence, deflection, and an expectation that I would absorb the impact without complaint.

What had been framed as leadership revealed itself to be little more than rhetoric. There was no protection for people doing the work, no reinforcement of standards, and no willingness to confront damaging behavior if it threatened the appearance of stability. Responsibility flowed downward, blame followed it, and support existed only as long as it was never tested.

After some time had passed, a management meeting was finally held. Managers from different teams were brought together, ostensibly to share updates and discuss ongoing projects. What should have been a forum for collaboration quickly turned into a public flogging. Senior leaders dominated the room through humiliation, intimidation, and outright bullying. Several had just returned from a visit to another station in the southeast and used it as a weapon, berating all of us by claiming the managers there were far better than we were. The message was blunt and unmistakable. Improve drastically or get out.

It went further than posturing. One experienced Program Manager was singled out, yelled at in front of the group, and told to start looking for another job. There was nothing constructive about it. No coaching. No context. No attempt to solve anything. It was pure performance, designed to assert dominance rather than drive improvement. That was when it finally clicked. This was not a system built to support managers or develop leaders. It was a system designed to break them, then blame them for the damage.

When I later raised concerns about the professionalism of the meeting to my manager, something I would come to recognize as a mistake in hindsight, there was no response at all. No discussion. No acknowledgment. Just silence. The shift was immediate and unmistakable. I was suddenly on the outside, alienated and marked as someone who had crossed an invisible line by questioning behavior that was never meant to be questioned. In a company that preached "Earn Trust," defined by listening attentively, speaking candidly, and treating others respectfully, the silence said everything. There was no listening. No candor. No respect. Trust was not broken in that moment. It was deliberately withheld.

Once I saw that clearly, there was no unseeing it. That was the moment the illusion finally collapsed. I was not failing to lead. I was being used as a buffer between avoidance upstream and the damage it created downstream, expected to absorb dysfunction quietly and accept the blame when it inevitably surfaced. Accountability only flowed downward. Support was conditional. Integrity was treated like a liability. What they called leadership was really just risk management by intimidation, a culture that protected itself by breaking the people closest to the work. Once I understood that, the question was no longer how to fix the team or survive the role. It was how long I was willing to stay inside a system that mistook erosion for strength and called it normal.

Standard Work Without Accountability

On paper, the PA Daily Standard of Work was rigorous, structured, and essential to the health of the network. Each PA was responsible for monitoring a defined group of more than sixty North American Sort Centers, conducting deep analysis to identify missort anomalies and trace them back to their root causes. The work demanded precision, consistency, and strict adherence to timelines, because even small misses created downstream consequences that rippled across operations, partners, and customer commitments.

The process was complex by design. Missort analysis involved more than a dozen discrete steps and required PAs to work across multiple internal tools to track package flow, isolate failure points, and validate data accuracy. Once an issue was confirmed, a Simple Issue Management (SIM) ticket documented the root cause and assigned ownership for resolution. Those escalations then moved to downstream partners like MMPM, which oversaw third-party logistics providers, and POST, the team coordinating with USPS, making accuracy at the front end nonnegotiable.

Everything was documented in Quip, which served as the system of record. Findings were logged, timelines tracked, and follow-ups monitored to ensure issues were not merely acknowledged, but actually resolved. PAs were expected to manage their assigned cases end to end, respond quickly to follow-up requests, and escalate unresolved or critical issues to me when necessary. When executed correctly, the process did more than fix isolated problems. It prevented repeat failures, protected delivery commitments, and reinforced trust across the network. The entire system depended on discipline and enforcement. Without those, the structure was meaningless.

That was the theory. In practice, the entire system unraveled the moment accountability was required.

As I continued building theTeam, I focused on establishing clear Standards of Work, enforcing accountability, and restoring basic discipline to a group that had gone without it for far too long. From the start, it was an uphill battle. The team had grown accustomed to minimal oversight and maximum leniency, and any attempt to introduce structure was met with resistance. Discipline felt intrusive. Accountability felt optional.

Many PAs openly resisted change. Some refused to adhere to standard work hours. Others missed deadlines or ignored escalation timelines altogether. When I pushed back, they bypassed me entirely, going straight to Cruella with complaints about "excessive workload" or my insistence on basic professional standards. Rather than addressing the behavior, that response validated it, reinforcing the idea that resistance was not only tolerated but rewarded.

Rather than supporting the expectations I had been asked to implement, Cruella actively undermined them. Each attempt to enforce standard work was met with reprimands directed at me, not the PA responsible for the miss. Accommodations were demanded that made adherence to timelines impossible, followed by criticism when the lack of results those same accommodations inevitably produced became visible. The message to the team was unmistakable. My authority was conditional. My decisions were reversible. Resistance worked. The dynamic did not drift into toxicity. It accelerated toward it.

Deadlines slipped. Critical issues lingered unresolved. Accountability continued to erode. Every corrective action triggered backlash instead of improvement. I was caught between a team unwilling to change and a manager determined to shield that resistance from consequence. The harder I pushed to stabilize the process, the more pressure came back at me from both directions.

The breaking point came during a particularly tense exchange when Cruella berated me in front of others for the team's performance. The criticism was not just unfair, it was completely detached from reality. I was being blamed for the very dysfunction that had been created, tolerated, and actively protected long before I arrived, and that continued to be reinforced in real time. After absorbing far more than I should have, I finally addressed it behind closed doors. "You set me up for failure by creating a toxic environment through your management." The words landed heavy, not because they were dramatic, but because they were true, and once spoken, there was no taking them back.

From that moment on, whatever remained of our working relationship collapsed. It was the point where obstruction stopped feeling passive and began to feel deliberate. Undermining that had once been subtle became harder to ignore. Conversations with upper leadership started reflecting narratives that barely resembled what was actually happening on the floor. Responsibility for the team's failures shifted quietly onto me, while the conditions that caused them went unexamined.

For the PAs, it meant continued freedom from accountability. For me, it meant fighting on two fronts, against apathy below and obstruction above. The workload was relentless. The stress constant. Support was effectively nonexistent. Instead of fostering improvement, the culture reinforced a system where blame was safer than honesty and appearances mattered more than outcomes.

Over time, it became clear that Cruella was not an anomaly. That role was a product of a culture that tolerated this behavior and, in some cases, rewarded it. Promises of support were offered easily and withdrawn quietly the moment real resistance appeared. Meetings that were meant to solve problems devolved into blame sessions. Strategy gave way to intimidation. Accountability flowed downward and stopped there.

Looking back, the environment did not thrive because of one toxic manager, but because the system allowed it to. The system embodied the culture, but did not invent it. The behavior was reinforced, protected, and normalized by a structure that shielded those above while grinding down the people closest to the work.

The scars from that period remain, not as resentment, but as clarity. A reminder that leadership is not measured by metrics alone. It is measured by what it tolerates, what it protects, and who it is willing to sacrifice to preserve the illusion of control.

That experience left me with a hard and lasting truth. Systems do not fail quietly. They fail because comfort is chosen over confrontation and optics over accountability. I had been asked to build quality inside a culture that resisted it at every turn, then punished anyone who refused to play along. The work taught me more than any title ever could. Quality is not created by dashboards or standard work alone. It lives or dies by the courage to enforce it. And when dysfunction is protected instead of corrected, the damage is never confined to process or performance. It lands squarely on the people still trying to do the job right.

When the Work Spoke for Itself

Without clear direction from above, I stopped waiting for it and charted my own course, starting with Weekly Business Reviews that brought impacted site leadership into the same room. Those WBRs quickly became a critical platform for collaboration, connecting me directly with leaders across Sort Centers, Fulfillment Centers, and Distribution Centers throughout North America, while also strengthening partnerships with downstream teams responsible for third-party logistics carriers and USPS operations. What began as an attempt to impose structure on an ambiguous role evolved into something far more effective, creating alignment, accountability, and trust across teams that rarely had the opportunity to work together in any meaningful way.

Those relationships proved essential. Through direct collaboration, we moved beyond surface-level assumptions and uncovered the real drivers behind missorts. One of the most consequential discoveries was how frequently mechanical failures were responsible for errors that had long been blamed on associates. At Sort Centers, robotics systems played a central role in moving and diverting packages along conveyor belts, so I led deep dives into the linear robotics systems tied to those operations and began identifying consistent patterns of scanner failure. Barcodes were being misread, partially obscured, or improperly aligned as packages moved at speed, creating downstream errors that no amount of coaching or discipline on the floor could ever fix.

Armed with data, I worked directly with national leadership and mechanical engineering teams to address those root causes, focusing on scanner positioning, calibration, and mechanical tolerances. These adjustments reduced error rates in ways manual intervention never could, delivering systemic fixes to systemic problems with immediate and measurable impact.

The work extended beyond Sort Centers. At Fulfillment Centers, I identified recurring issues where packages were being incorrectly diverted to UPS instead of their intended carriers, creating delivery delays and unnecessary cost that quietly eroded efficiency across the network. By digging into routing logic, handoff points, and carrier assignment data, I proposed targeted corrections that reduced misroutes and stabilized last-mile flow.

Taken together, these efforts did more than resolve isolated problems. They laid the groundwork for sustained improvement across the network, with insights from deep dives informing broader changes in robotic configuration, barcode handling standards, and operational decision-making. Over time, the missort cost needle moved in the right direction, saving Amazon millions of dollars and proving that real results were possible when ownership, collaboration, and data were prioritized, even without clear top-down guidance.

None of that stopped Cruella. No matter how much progress was made, resistance followed, with negativity and manipulation casting a shadow over every win. And yet, despite that interference, something unexpected began to happen as a small group of PAs started to believe in the work, leaned in, took ownership, and shifted from passive participants into genuine contributors. Watching that transformation unfold became one of the few bright spots during that period, a reminder of what people are capable of when given clarity, trust, and purpose.

Those moments mattered, because they reinforced why the effort was worth it, even as they never lasted long. For every success, there was an attempt to diminish it through nitpicking inconsequential details, discrediting the team's work, or quietly redirecting credit toward someone who had actively obstructed progress. Wins were never acknowledged without qualification, and momentum itself was treated as a threat rather than an asset.

It became clear that success would never be allowed without resistance. The grip on the culture that had been created was firm, and its influence extended well beyond the immediate team. Narratives were shaped carefully, relationships leveraged strategically, and undermining others was not merely tolerated but rewarded. The work continued to move forward, but the environment refused to follow.

In the end, the contrast was impossible to ignore. When I was free to focus on the work, progress followed. Costs dropped. Errors declined. Relationships strengthened. People stepped up. When interference returned, progress stalled, not because the solutions were flawed, but because control and ego mattered more than outcomes. That realization settled in quietly but completely. Results were never the problem. Culture was. And once I understood that, the work stopped being about fixing systems and started being about recognizing when a system could no longer be fixed from the inside.

I stayed because, for a long time, the system worked for me. I was rewarded for fixing problems. I was promoted when things improved. I succeeded by Amazon's own definitions, trusted with larger scopes, harder assignments, and roles meant to stabilize what was already breaking. Each step forward reinforced the belief that effort, competence, and integrity still mattered, that progress inside the machine was possible if you understood it well enough. That success is precisely why I remained long enough to see the pattern clearly. The farther I advanced, the wider the view became, and the harder it was to ignore that the same dynamics repeated regardless of role, station, or leadership layer. What initially felt like isolated failures revealed themselves as design features, and the rewards that once kept me invested became the reason I could no longer pretend the system was merely flawed rather than fundamentally misaligned.

The Point of No Return

The tipping point came during a meeting with a leader who embodied everything toxic I had encountered during my years at Amazon. In the moment, I thought of him as one of the most damaging leaders I had ever encountered at Amazon, and even with distance, that assessment still holds. What made it revealing was not just the incompetence, but the confidence behind it. It would have been almost comical if it had not perfectly exposed the kind of leadership the system produced. His failure went far beyond a lack of understanding of my role. He lacked curiosity, critical thinking, and the ability to listen or engage with complex work. What he did possess was absolute certainty in his own ignorance.

By that point in my career, the pattern was familiar. Different titles. Different orgs. Same behavior. Leaders who advanced not because they understood the business, but because they understood politics. Loyalty mattered more than truth. Alignment mattered more than results. Questioning the narrative was treated as a liability. This person was not an anomaly. He was the logical outcome of a system that rewarded obedience over competence and confidence over understanding.

He walked into the meeting already poisoned by Cruella's version of events, armed with talking points and a dangerously shallow grasp of the work itself. He did not ask a single clarifying question. Instead, he launched into an aggressive attack on everything I had built. He criticized results without knowing the metrics. He questioned methods without understanding the process. He challenged intent as if fixing a broken system were somehow self-serving. Every accusation was secondhand. Every conclusion was borrowed. He repeated her narrative with conviction, mistaking allegiance for leadership.

What made it infuriating was not just the stupidity. It was the certainty. The smug confidence with which real outcomes were dismissed in favor of convenient fiction. Sitting there, watching him tear into work he could not possibly comprehend, I saw the culture laid bare. This was an environment where incompetence did not just survive. It flourished. Leaders like him, ignorant, compliant, and incapable of independent thought, could rise as long as they aligned themselves with the right personalities.

I had seen this dynamic play out for years. Strong operators were sidelined, thoughtful leaders burned out, and people who asked hard questions were labeled difficult. Meanwhile, manipulators were protected, bullies were promoted, and those willing to enforce dysfunction without protest were elevated. Results mattered only when they supported the existing story. When they did not, they became something to challenge, minimize, or attack.

In that moment, something crystallized. It did not matter how much value I delivered, how many problems I solved, or how much money I saved the company. In a system this broken, success was not simply unrecognized. It could become threatening. Cruella's sabotage, combined with the blind complicity of leaders like this one, made the reality unmistakable. There was no meaningful path forward for anyone unwilling to play along. The system was not built to reward integrity or reform. It was built to protect itself.

That meeting became my breaking point, and the decision came with absolute clarity as I sat across from one of the most disconnected leaders I had encountered. I was done. Not angry-done or burned-out-done, but finished. Walking away was not dramatic or reckless. It was necessary, the only way to protect my sanity, my self-respect, and any future worth having outside a company so committed to defending its own dysfunction.

For the first time in a long while, the path forward was obvious, and it did not include Amazon.

That realization brought an unexpected calm. Once I accepted that the system was never going to change, the confusion, anger, and self-doubt fell away. I no longer needed to explain myself, defend my work, or bend my values to survive in an environment that treated improvement as a threat. Leaving was not failure. It was clarity.

When I first stepped into Amazon's machine, I was drawn to its precision, its scale, and its ability to turn complexity into motion. I believed that if you understood the system deeply enough, if you respected the data and the people behind it, you could make it better. For a long time, that belief carried me. It drove me to solve problems, build systems, and push for outcomes that mattered.

But machines do not change themselves. They respond only to the incentives built into them. And this one had been engineered to prioritize speed over reflection, obedience over curiosity, and control over care. Innovation flourished. Humanity did not. The cost I once wondered about was no longer abstract. I had lived it.

Amazon had taken everything it was going to take from me, and I had learned everything it was capable of teaching. What remained was mine to protect. Walking away was not an act of defeat, but of preservation. I did not leave because I could not survive inside the machine. I left because I finally understood what surviving there required, and what it would cost to keep doing it.

This book began inside the engine room of a company that reshaped the modern world. It ends with a simple truth. Systems this powerful do not break by accident. They break people quietly, predictably, and often invisibly, until someone decides to step out of the gears. This is what it's like to live and work inside the machine. And this is what it took to walk away from it.

Controlling the Narrative

Amazon's ability to dominate industries appeared to me to go far beyond logistics and e-commerce. It also extends to shaping how its internal operations are perceived. With vast resources and enormous influence, the company has become highly effective at presenting a carefully curated public image built around innovation, efficiency, and customer obsession. The version of Amazon most people see is deliberate, consistent, and polished. Understanding what happens inside its facilities requires looking beyond that surface and listening closely to the people whose experiences rarely make it into official narratives.

Over the years, I learned that many of the most troubling events inside the network never become public stories. Not because they are insignificant, but because they are absorbed internally, documented quietly, and then allowed to disappear. Associates talk. Managers talk. Patterns emerge. But those stories tend to stay contained, circulating among employees rather than reaching the outside world. The result is a widening gap between the company's external messaging and the internal reality experienced by people on the floor.

One account that stayed with me involved a violent incident at a West Coast station. According to multiple people familiar with the situation, an associate was assaulted by coworkers while following guidance that emphasized avoiding physical confrontation under any circumstances. The details, as they were described to me, were disturbing not only because of the violence itself, but because of how the situation was handled afterward. The incident was treated as an internal matter, as it was described to me. It never became a broader conversation about safety, policy, or accountability. For those who knew about it, the silence that followed was as unsettling as the event itself.

What struck me most was the contrast between those accounts and the company's public posture. Amazon regularly emphasizes its commitment to workplace safety, yet the experiences shared internally suggested that policies designed to reduce liability could leave people vulnerable in moments when protection mattered most. Whether intentional or not, the effect was the same. The people closest to the work felt unseen, unheard, and expendable.

I heard similar stories elsewhere. At another station, an act of extreme workplace violence involving a manager and a delivery associate was described to me in vivid detail by people who were close to the situation. The manager survived, narrowly, and the incident left lasting physical and psychological damage. What followed, according to those accounts, was a familiar pattern. The event was handled internally. Information was tightly controlled, according to those accounts. Outside of the immediate circle of those affected, the incident faded quickly from view. For the people who lived through it, the absence of transparency reinforced a hard lesson about how uncomfortable truths are managed.

Violence was not the only area where narrative control was evident. I also watched how Amazon responded to collective worker action. As unionization efforts gained traction in various fulfillment centers across the U.S. and Canada, the company's public statements emphasized flexibility, opportunity, and employee choice. Internally, however, the outcomes often involved large-scale operational changes that displaced entire workforces. Whether framed as business decisions or strategic realignments, the effect on workers was immediate and destabilizing. Addressing the underlying concerns was rarely part of the conversation.

Beyond labor issues, Amazon has invested heavily in shaping the regulatory and public environment in which it operates. Public records show that the company has spent tens of millions of dollars on lobbying efforts, engaging lawmakers and agencies on issues ranging from consumer privacy to environmental regulation. These efforts are legal and common among corporations of Amazon's size, but their scale matters.

Influence at that level shapes which conversations gain traction and which quietly stall.

There are also well-documented concerns around surveillance and monitoring inside Amazon facilities. The company has patented technologies capable of tracking worker movements with extreme precision and has faced criticism for monitoring employee communications and online activity. Whether framed as efficiency or safety measures, these practices contribute to a culture where speaking openly about conditions carries perceived risk. Silence, once again, became the safer option.

Taken together, these patterns reveal something important. Amazon's power does not come solely from its technology or logistics network. It comes, in my experience, from its ability to manage perception, contain uncomfortable stories, and keep control of the narrative surrounding its operations. None of this requires explicit malice, as I came to understand it. Systems like this function, in my experience, through incentives, risk avoidance, and silence as much as through direct action.

For me, understanding that dynamic helped explain why so many internal problems felt invisible from the outside. It explained why people suffered quietly, why accountability was so difficult to pin down, and why meaningful change was so slow. The public story and the internal reality often existed far apart, and bridging that gap required listening to voices that rarely made it past the warehouse walls.

This chapter is not about exposing secrets. It is about acknowledging a pattern. When scale becomes power and control becomes priority, truth becomes negotiable. And in an organization as vast as Amazon, what goes unseen can matter just as much as what is carefully put on display.

What It Cost, and What Remained

Looking back at my time at Amazon, I no longer see it as just a career. I see it as a prolonged test of endurance, resilience, and survival. I walked into the company ambitious and motivated, believing I could build something meaningful inside one of the most powerful logistics machines in the world. I bought into the promise that hard work mattered, that data-driven decision making and operational excellence created space for people who were willing to do the work and do it well. In many ways, that part was true. I learned the systems, improved processes, led teams through controlled chaos, and contributed to operations designed to run with ruthless precision.

But those accomplishments came at a cost I did not fully understand at the time.

Alongside the work was a deeply corrosive leadership culture, one that rewarded loyalty over competence, punished accountability, and protected mediocrity at the highest levels. I watched managers with no business leading people wield unchecked power over careers. I saw how those who questioned broken processes, challenged bad decisions, or simply tried to enforce basic standards were sidelined, labeled as problems, or pushed out entirely. Amazon was not just a place of opportunity. It was a place where success often depended less on ability and more on how well you learned to navigate politics and stay quiet when things went wrong.

I endured constant sabotage from leaders who treated authority as a personal weapon rather than a responsibility to others. I watched ineffective managers rise because they defended the system instead of improving it. Undermining, gaslighting, and bullying were not exceptions. They were normalized. Leadership either ignored it or actively participated. And still, I kept going. I refused to back down, even when the pressure became relentless.

It took nearly a year after leaving Amazon for the damage to truly surface. My wife would tell me that I screamed in my sleep. There were mornings when I got dressed for work while still asleep, operating on muscle memory and stress alone. The physical toll was just as real. The injuries to my ankles never fully went away. I still see specialists. I still attend physical therapy twice a week to manage the damage. During that time, I coped the wrong way. I drank too much. I ate poorly. I used food and alcohol to numb stress I did not yet know how to process.

I am proud to say that chapter is closed. I no longer drink. I made fitness a priority and lost more than fifty-five pounds. I rebuilt my health deliberately and with intention. Amazon took a real psychological and physical toll on me, but it did not get to keep the ending. I moved past it. I reclaimed my body, my sleep, and my sense of peace. I hope the good people I worked alongside were able to do the same.

Through sheer determination, I proved I was more than the system's attempt to break me. I delivered results even when I was set up to fail. I built processes that worked. I developed teams that responded to real leadership. I refused to let a toxic environment dictate my value or erase my integrity. I saw firsthand the brutal cost of Amazon's obsession with efficiency, how it squeezed everything it could from people before discarding them like replaceable parts. And yet, in that environment, I also discovered my own strength.

Amazon did not define me. The leaders who tried to undermine me did not define me. What defines me is that I entered one of the most unforgiving environments imaginable and held my ground. I adapted. I succeeded. And when it became clear that no amount of competence could fix a system unwilling to change, I left on my own terms.

What I carry with me now are hard-earned lessons about what leadership actually means. Leadership is not fear. It is not power. It is not numbers on a spreadsheet. It is the impact you have on people, the standards you protect, and the integrity you refuse to surrender, even when the environment around you is toxic. Real leadership builds people instead of grinding them down. It creates space for honesty, accountability, and growth.

Amazon will continue operating the way it always has. Relentless. Impersonal. Efficient at almost any cost. I walked away with something far more valuable. I faced every challenge head-on. I refused to become part of the dysfunction. And in the end, I came out stronger, clearer, and fully in control of my own future.

That is what I carry forward. Not resentment, not regret, but clarity. No toxic leader, broken system, or flawed culture gets to define what I built, what I learned, or who I became.

The Cost of Amazon's Future

With all the success Amazon has already achieved, the more interesting question is no longer what it built, but what it becomes next. The company has fundamentally reshaped global commerce, logistics, and cloud computing, yet its ambitions extend far beyond those achievements. Amazon's next phase is centered on deeper automation, expanded artificial intelligence, and tighter integration into everyday life. Whether through autonomous delivery, predictive commerce, or entry into entirely new industries, Amazon is positioning itself not just to compete, but to redefine the boundaries of modern capitalism.

One of the most visible examples of this push is Amazon Prime Air, the company's drone-based delivery initiative. Currently being tested in Lockeford, California, and College Station, Texas, the program aims to move small packages directly to customers within hours. The stated goal is speed and convenience, but the strategic objective is clear: reducing reliance on traditional last-mile delivery networks and consolidating control inside Amazon's own ecosystem. Alongside drone delivery, Amazon has also experimented with autonomous ground delivery through Scout, signaling a long-term vision in which human involvement in last-mile logistics is minimized wherever possible.

Beyond delivery, Amazon continues to expand through strategic partnerships and acquisitions that extend its reach into adjacent markets. Its collaboration with Kohl's, allowing customers to return Amazon packages in-store, increased convenience for customers while driving foot traffic to struggling brick-and-mortar retail. Its partnership with Grubhub reflects an effort to strengthen its position in food and grocery delivery, a sector already dominated by companies like Uber Eats and DoorDash. With Amazon's logistics infrastructure, data capabilities, and Prime ecosystem, even incremental moves into these spaces carry the potential for major disruption.

Amazon's acquisition strategy has been just as aggressive. The $8.5 billion purchase of MGM Studios marked a decisive move into entertainment ownership, giving Amazon access to iconic franchises and valuable intellectual property to bolster Prime Video. Its nearly $4 billion acquisition of One Medical signaled an even more consequential shift into healthcare. Combined with existing services like PillPack and health-related integrations through Alexa, Amazon is steadily embedding itself into some of the most personal and sensitive areas of consumers' lives.

Artificial intelligence and automation sit at the center of all of these efforts. AI-driven customer service, predictive purchasing algorithms, and recommendation systems are designed to remove friction from the shopping experience and anticipate consumer needs before they are explicitly expressed. Inside Amazon's fulfillment network, robotic arms, automated storage systems, and machine-learning-driven inventory management continue to replace human labor in the name of speed and efficiency. Whether these advancements ultimately improve working conditions or simply accelerate workforce displacement remains an open and unresolved question.

For all its innovation, Amazon's future carries a familiar shadow. The same relentless pursuit of efficiency that fuels its growth has consistently come at a cost to its workforce. In warehouses and delivery networks, workers face intense productivity demands, constant surveillance, and injury rates that remain higher than industry averages. Data, metrics, and optimization drive decision-making, often reducing human labor to a variable to be monitored, maximized, and replaced when no longer optimal.

Amazon's next chapter will undoubtedly involve continued expansion, deeper technological integration, and increased market dominance. But its long-term legacy will be defined by more than innovation alone. As the company pushes further into logistics, healthcare, entertainment, and daily life, the unresolved tension between progress and accountability grows harder to ignore. The balance Amazon chooses between efficiency and humanity will determine whether its future is remembered as a story of transformative innovation or unchecked corporate power.

One thing is certain. Amazon's story is far from finished.

Epilogue: To Those Still Inside the Machine

If you are still inside Amazon as you read this, this epilogue is for you. Not to persuade you to leave. Not to tell you what you should do. Only to acknowledge what you already know but may not always be allowed to say out loud.

You are not imagining the pressure. You are not weak for feeling worn down by it. And you are not failing simply because the system keeps moving the goalposts faster than any human can reasonably adapt. Amazon is an extraordinary machine. It is capable of breathtaking scale, speed, and innovation. It can solve problems most organizations would never even attempt. It can also demand more than it admits, normalize exhaustion, and quietly teach people to doubt themselves when the machine resists being questioned.

If you care about doing the work well, you have likely felt the tension. The conflict between building something that actually works and navigating incentives that reward appearance over substance. The discomfort of seeing problems clearly while being encouraged, subtly or directly, not to push too hard. The exhaustion that comes from carrying responsibility without authority, ownership without protection. None of that is accidental. Systems like this do not rely on cruelty to function, from what I saw. They relied on silence, compliance, and the hope that people would adapt just enough to keep everything moving.

There are good people inside Amazon. Talented, principled, deeply committed people. Many of them are still there because they believe the work matters, because they are supporting families, because they are learning, or because they are trying to change things from the inside. All of those reasons are valid. Staying does not make you complicit. Leaving does not make you brave. What matters is honesty, especially with yourself.

Pay attention to what the system rewards and what it quietly punishes. Notice when integrity costs more than it should. Notice when asking good questions becomes risky. Notice when success feels threatening rather than celebrated. Those signals matter. They are not personal failures. They are information.

My story ended with leaving, but that does not have to be yours. This book is not a prescription. It is a mirror. Use it to orient yourself. Protect your health. Protect your values. Build skills that travel with you. Invest in relationships that see you as more than a metric. And remember that your worth is not defined by how well you endure pressure inside any one organization.

Machines are built to keep running. People are not. Whatever choice you make, make it consciously, not by default. The moment you stop believing that the cost is invisible is the moment you begin to regain control.

Take care of yourself. The work will always ask for more. The question is what you are willing to give, and what you are no longer willing to lose.

Author's Note

This book was not written for revenge. It was written for accuracy.

Everything you have read reflects my lived experience inside Amazon's operational machine, shaped by years of work across multiple roles, teams, and leadership layers. Some details have been generalized, names, locations changed and timelines compressed to protect individuals who did not choose to be part of this story. The substance remains intact.

What I describe is not, in my experience, a collection of isolated incidents. It is a culture reinforced through incentives, silence, and avoidance. It is a system that functions extraordinarily well at scale, yet too often fails the people closest to the work. These failures felt structural.

Writing this required distance and restraint. I waited until I could tell the story without exaggeration and without needing validation from anyone still operating inside the machine. I am not asking the reader to agree with every conclusion. I am asking them to consider the cost of what works, who pays it, and why those costs are so rarely acknowledged.

This is not the final word on Amazon, only one honest account from my vantage point inside its gears. The real question is whether I would do this all over again. Minus the worst of the unhinged leadership dysfunction, the answer is yes. Amazon fulfilled my strong need to learn, lead, change lives for the better, implement new processes, and adapt to a constantly changing environment. It was not a place for the faint of heart, because it constantly challenged your judgment, stamina, patience, and ability to keep moving when the ground underneath you was always shifting. That is why this account matters to me. It is not written to settle every argument about Amazon, but to show what it felt like to work inside a system that could develop people, test them, use them, and sometimes break them, all at the same time.

All experiences, judgments, interpretations, and conclusions in this book are my own. AI-assisted editing tools were used to support copyediting and line-level revision, including grammar, spelling, and sentence structure. They did not generate the experiences described here, the underlying judgments, or the structural argument.

About the Author

Rad Stephens spent more than three decades working inside large-scale operational environments, building and managing the systems that move work, people, and decisions at scale. He began on the floor in distribution and sort centers and advanced into leadership roles responsible for multi-site logistics operations, where he oversaw performance, staffing, and the data systems used to measure both.

At Amazon, he managed operations across multiple facilities and led initiatives focused on network quality, performance metrics, and process design. His work placed him at the intersection of frontline execution and executive decision-making, where he observed firsthand how data, automation, and emerging A.I. systems began to shape not just outcomes, but authority itself.

Stephens' writing draws directly from that experience. His work focuses on the human consequences of scale, the quiet transfer of decision-making into systems, and the gap between what organizations measure and what the people inside them actually know. He is the author of *Amazon Unfiltered: Power, Performance, and the Human Cost of Scale* and *The System Is the Boss*, both of which examine modern organizational life from the inside out.

www.ingramcontent.com/pod-product-compliance
Lightning Source LLC
LaVergne TN
LVHW090352240926
848801LV00001B/17